Modeling, Simulation, and Optimization of Supply Chains

Modeling, Simulation, and Optimization of Supply Chains

A Continuous Approach

Ciro D'Apice
University of Salerno
Fisciano, Italy

Simone Göttlich
University of Kaiserslautern
Kaiserslautern, Germany

Michael Herty
RWTH Aachen University
Aachen, Germany

Benedetto Piccoli
Rutgers University
Camden, New Jersey
IAC-CNR
Rome, Italy

siam®
Society for Industrial and Applied Mathematics
Philadelphia

10 9 8 7 6 5 4 3 2 1

This work was financially supported by Deutsche Forschungsgemeinschaft (DFG), Deutscher Akademischer Austauschdienst (DAAD), RWTH Aachen Seed Funds, the Istituto Nazionale d'Alta Matematica "Francesco Severi" (INdAM) through the project "Traffic flows and optimization on complex networks," and the EU Network of Excellence "HYCON."

Library of Congress Cataloging-in-Publication Data

Modeling, simulation, and optimization of supply chains : a continuous approach / Ciro D'Apice ... [et al.].

p. cm.

Includes bibliographical references and index.

ISBN 978-0-898717-00-6

1. Business logistics—Mathematical models. I. D'Apice, Ciro, 1967-

HD38.5.M63 2010

658.701'1—dc22

2009052937

To my father Domenico
(C. D'Apice)

To my husband Joachim
(S. Göttlich)

To my son Edoardo
(B. Piccoli)

Contents

Preface

This book is devoted to primarily continuous models for a special class of supply chains often called production or supply networks. The aim is to present a mathematical description of different phenomena appearing in planning and managing supply chains. We address both the mathematical modeling as well as techniques for simulation and optimization purposes.

The problem of a continuous description of supply chains and production networks dates back to the early 60's and started with the work of [8, 30]. Significantly, the models were proposed in particular for large volume production on complex networks where a discrete description might fail. Since then, many methods and ideas have been developed concerning the modeling of different features of supply chains, including the efficient simulation and the optimization of product flows among suppliers and customers. In recent years continuous and homogeneous product flow models have been introduced, for example, in [2, 13, 23, 26, 27, 28, 29, 37, 41, 42]. These models have been built in close connection to other transport problems like vehicular traffic flow and queuing theory. Hence, this suggests that the obtained models should be given by partial differential equations for the product flow, similar to those of gas dynamics. Depending on the problem at hand, these equations are possibly accompanied by ordinary differential equations describing the load of inventories. Also some optimization techniques have been proposed in order to answer questions arising in supply chain planning [31, 50, 57].

Starting from a network formulation, we derive equations for a continuous description of homogeneous product flows. The derivation is based on first principles, but the final equations are closely related to discrete event simulations of supply chains. Additionally, we present extensions to include more realistic phenomena. Such extensions consist of systems of partial differential equations or coupled partial and ordinary differential equations. The book surveys the underlying fundamentals and provides evolved mathematical techniques for simulation and efficient optimization of the presented models.

The book is suitable for researchers and students in mathematics, physics, engineering, economics, and related disciplines. Basically, the book could also be used as lecture notes for a course in supply chain theory. Furthermore, interested readers can find many suggestions and open problems for future research.

At first, in Chapter 1, we give some elementary examples which serve as motivation for the modeling approaches following, and then we report mathematical preliminaries about conservation laws and numerical schemes for their discretization. Chapters 2–5 contain a complete and elementary description of the state-of-the-art theory of continuous supply chain models where also similarities and comparisons with already existing models are given. Optimization problems and the corresponding solution techniques are the content of Chapter 6. In particular, the interplay between discrete and continuous optimization problems is addressed. Numerical discretization issues and computational results in Chapter 7 complete this book.

We would like to thank the staff at SIAM for their expert guidance and assistance during this project. Special thanks go to Elizabeth Greenspan for her friendly and helpful manner in answering editorial aspects. Furthermore, we are very grateful to our colleagues Dieter Armbruster, Gabriella Bretti, Pierre Degond, Axel Klar, Claus Kirchner, Sebastian Kühn, Rosanna Manzo, and Christian Ringhofer, whose collaboration, interesting discussions, and valuable advices influenced the material presented within this book. Finally, we also wish to express our gratitude to our families for their patience and permanent encouragement. This work was financially supported by Deutsche Forschungsgemeinschaft (DFG), Deutscher Akademischer Austauschdienst (DAAD), RWTH Aachen Seed Funds, the Istituto Nazionale d'Alta Matematica "Francesco Severi" (INdAM) through the project "Traffic flows and optimization on complex networks," and the EU Network of Excellence "HYCON."

September 2009
Salerno, Italy
Kaiserslautern, Germany
Aachen, Germany
Rome, Italy

Ciro D'Apice
Simone Göttlich
Michael Herty
Benedetto Piccoli

Chapter 1

Introduction

A supply chain consists of suppliers, manufacturers, warehouses, and stores where parts are produced and distributed among different production facilities. Mathematical models are used to monitor cost-efficient distribution of parts and to measure current business processes. Naturally, depending on the scale, these models are characterized by several approaches which are either discrete (discrete event simulations) or continuous (differential equations). The main difference between these two mathematical concepts is the description of parts as individuals at discrete time instances or as a dynamic flow.

Simulations in general represent a powerful computing technique to analyze manufacturing systems while performing numerical experiments of the models. In case of discrete event simulations the evolution of the system is viewed as a sequence of significant changes in time, also called events, for each part separately. For instance, consider a supply chain consisting of numerous consecutive facilities where parts arrive, get processed, and depart when their service is completed. Informally, this means that the transportation of parts from one production step to another characterizes dynamic events that can be easily evaluated using performance measures like the number of parts in the system, the individual waiting times, and so forth. Definitely, discrete event simulations serve as the finest level of description of interacting part-based systems but with the drawback of exponentially increasing computing times for large-scale systems, i.e., up to 10^3 and more suppliers.

An alternative modeling approach which remedies the computational aspect is differential equations. In contrast to the discrete event simulation, averaged quantities in case of large quantity production predict the time evolution of parts and include the dynamics inside the different production steps. To derive accurate continuous models the overall modeling goal is to transfer as much of the detailed and complex discrete model to the continuous level. This will be achieved regarding dynamic flows, i.e., parts per time unit, instead of individual parts. Since numerical schemes for differential equations allow for fast simulation times, supply chain problems with multiple manufacturers and thousands of parts are solved very cost effectively. Figure 1.1 gives an idea of how computing times evolve when sample examples depending on 20.000 or 200.000 parts, respectively, are simulated. Keep in mind that partial differential equations, a special choice of differential equations, will always provide valid results if the size of the underlying problem is large scaled.

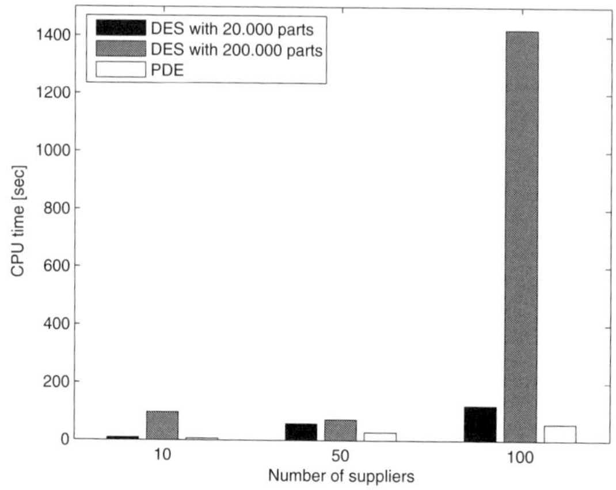

Figure 1.1. *The efficiency of using partial differential equations (PDE) in comparison with discrete event simulations (DES) is depicted. Three different scenarios of different complexity varying the number of suppliers and parts as well are simulated over* 400 *time periods.*

In many applications the simulation and prediction of production systems is only one important purpose. The formulation of optimization problems for supply chain management is an immediate consequence of performing successful simulations. Nowadays, there exist a variety of sophisticated models and adapted mathematical methods to find the optimal load balance on the interconnections between different entities. Typically, such problems are mainly tackled by methods of discrete optimization, linear and mixed-integer programming methods or models based on differential equations. Simpler models based on linear programs often neglect the time dependence of processes as well as nonlinear cost effects; however, they are applicable to large-scale networks. On the other hand, advanced optimization procedures recover the whole dynamics of the system while adjusting parameters in the models in an optimal way. Concerning manufacturing problems this may include the question of optimal processing parameters, minimizing inventories to reduce costs or to ensure fully loaded production lines.

Mathematically, supply chain optimization problems are given by either the minimization or maximization of an objective function (alternatively cost functional) satisfying flow conservation and capacity constraints. A relevant example would be the optimal routing of parts through a network such that inventory costs are minimized. This situation normally occurs whenever it is possible to distribute parts among several capacity-limited suppliers; cf. Figure 1.2.

Here, the rate $A^{1,2}$ denotes the percentage of parts coming from supplier e^1 continuing on supplier e^2, and equivalently $A^{2,6}$ determines the percentage of parts coming from supplier e^2 continuing on supplier e^6. The network allows for different capacity restrictions of suppliers and the possibility to store surplus material in queues q^e.

A bottleneck situation emerges from supplier e^6, which means that this capacity is the lowest in the complete network. The visualization of all objective function values

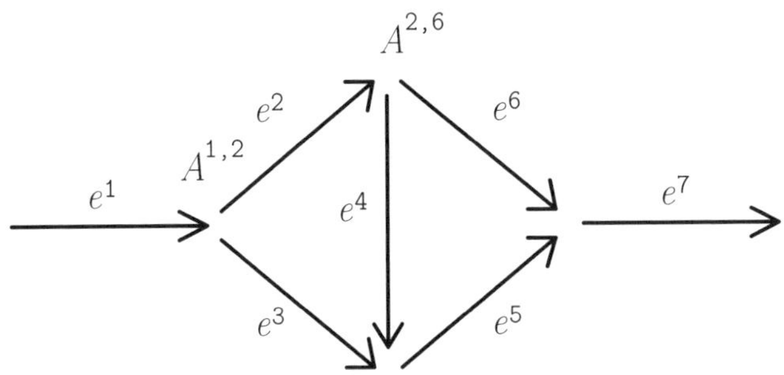

Figure 1.2. *A network consisting of seven suppliers and exactly two critical points $A^{1,2}$ and $A^{2,6}$ where the dispatch of parts obviously may be influenced.*

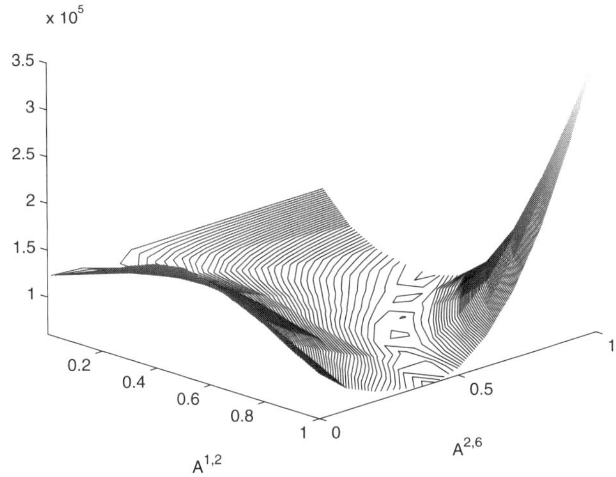

Figure 1.3. *Objective function values evaluated at all possible combinations of $A^{1,2}$ and $A^{2,6}$. As can be seen in the plot, the minimal objective value is reached at $A^{1,2} = 1$ and $A^{2,6} = 0.4$.*

in Figure 1.3 shows a rather naive way to choose the best value from a set of available alternatives. Nevertheless, we observe that the objective function has a unique minimum and steep gradients as $A^{2,6} \to 1$, due to the capacity bottleneck of supplier e^6, which implies an increasing inventory if filled by parts from supplier e^2. More evolved mathematical optimization methods are, for example, mixed-integer problems including dynamics which are consistent with the underlying detailed description by partial differential equations and/or particle/discrete event models. For the optimization of continuous supply chain models one could proceed in a different way. In the differential equation framework a natural approach would be to use an optimization procedure based on the Lagrange principle where the original constrained model is reformulated as an unconstrained one.

Altogether, the objective of this book will be to highlight the different ways of modeling supply chains and to emphasize their application for simulation and optimization purposes as well. In particular, similarities of the models will be discussed, and evolved mathematical methods governing the dynamic behavior of processes will be developed and analyzed. A major focus is the description of the underlying dynamics using ordinary and partial differential equations which may fit best in the context of mass production. Moreover, simulation and optimization procedures for these dynamical networks can be derived in a straightforward way combining already existing and innovative procedures.

Chapter 2

Mathematical Preliminaries

2.1 Introduction to Conservation Laws

Some of the models for supply chains we present in this book are based either on scalar or systems of hyperbolic conservation laws. A complete theory of hyperbolic systems of conservation laws is beyond the aim of the book. The theory of conservation laws is extensively described in many books, such as Bressan [10], Dafermos [21], Smoller [70], or Holden and Risebro [54]. Important results have been established in the past years and briefly revised some important results: the general solution to the Riemann problem for a strictly hyperbolic system of conservation laws was first obtained by Lax. The first proof of global existence for weak entropic solution appeared in the seminal paper by Glimm [34]. It is based on a construction of approximate solutions generated by Riemann problems with a randomly restarting procedure. For the scalar case, there is another proof, based on piecewise constant approximations, for the existence of an entropy admissible solution. This method is due to Dafermos [21]. The wave-front tracking method was first introduced by Di Perna, and then it was extended by Bressan [10]. Uniqueness and Lipschitz continuous dependence of solutions to scalar conservation laws (in many space variables) were first obtained by Kruzkov using the special entropies. The first proof of uniqueness for systems was obtained in 1996 by Bressan, Crasta, and Piccoli [11]. The proof was much simplified using the Bressan–Liu–Yang functionals; see [12].

In this section we give some basic preliminaries about systems of conservation laws.

Conservation Laws

A system of conservation laws in one space dimension can be written in the form

$$u_t + f(u)_x = 0, \tag{2.1}$$

where $u : [0,+\infty[\times\mathbb{R} \to \mathbb{R}^n$ is the "conserved quantity" and $f : \mathbb{R}^n \to \mathbb{R}^n$ is the flux.

Indeed, if we integrate (2.1) on an arbitrary space interval $[a,b]$, then

$$\frac{d}{dt}\int_a^b u(t,x)dx = -\int_a^b f(u(t,x))_x dx = f(u(t,a)) - f(u(t,b)),$$

and so the number of u in any interval $[a,b]$ varies according to the quantity of u entering and exiting at $x=a$ and $x=b$.

We always assume f to be smooth; thus, if u is a smooth function, then (2.1) can be rewritten in the quasi-linear form

$$u_t + A(u)u_x = 0, \tag{2.2}$$

where $A(u)$ is the Jacobian matrix of f at u.

Definition 2.1. *The system* (2.2) *is said to be hyperbolic if, for every $u \in \mathbb{R}^n$, all the eigenvalues of the matrix $A(u)$ are real. Moreover,* (2.2) *is said to be strictly hyperbolic if it is hyperbolic and if, for every $u \in \mathbb{R}^n$, the eigenvalues of the matrix $A(u)$ are all distinct.*

Remark 2.1.1. *It is clear that* (2.1) *and* (2.2) *are completely equivalent for smooth solutions. If instead u has a jump, the quasi-linear equation* (2.2) *is in general not well defined, since there is a product of a discontinuous function $A(u)$ with a Dirac measure. A notion of solution in the case of discontinuous functions is given in this section.*

If $n=1$ so u takes values in $\mathbb{R}$ and $f:\mathbb{R}\to\mathbb{R}$, then (2.1) is a single equation. In this case we say that (2.1) is a scalar equation. If $n>1$, then (2.1) is a system of n equations of conservation laws. Indeed if $u=(u_1,\dots,u_n)$ and $f=(f_1,\dots,f_n)$, then (2.1) can be written in the form

$$\begin{cases} \partial_t u_1 + \partial_x f_1(u) = 0, \\ \vdots \\ \partial_t u_n + \partial_x f_n(u) = 0. \end{cases}$$

Weak Solutions

A standard fact for the nonlinear system (2.1) is that classical solutions may not exist for some positive time, even if the initial datum is smooth. Let us consider, for example, the scalar Burgers equation

$$u_t + uu_x = 0,$$

with the initial condition $u(0,x) = u_0(x) = \frac{1}{1+x^2}$. One shows that the solution $u(t,x)$ to this Cauchy problem must be constant along the lines

$$t \mapsto \left(t, x + \frac{t}{1+x^2}\right).$$

For t sufficiently small ($t < \frac{8}{\sqrt{27}}$) these lines do not intersect together, and so the solution is classical, but at $t = \frac{8}{\sqrt{27}}$ the characteristics intersect together and a classical solution, i.e., differentiable solution, cannot exist for $t \geq \frac{8}{\sqrt{27}}$; see Figure 2.1.

Hence we must deal with weak solutions.

Definition 2.2. *Fix $u_0 \in L^1_{loc}(\mathbb{R};\mathbb{R}^n)$ and $T>0$. A function $u:[0,T]\times\mathbb{R}\to\mathbb{R}^n$ is a weak solution to the Cauchy problem*

$$\begin{cases} u_t + f(u)_x = 0, \\ u(0,x) = u_0(x) \end{cases} \tag{2.3}$$

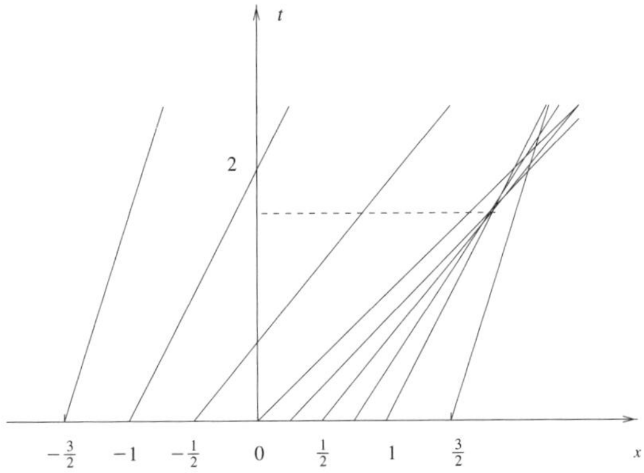

Figure 2.1. *The characteristic curves for the Burgers equation in the (t,x)-plane.*

if u is continuous as a function from $[0,T]$ into L^1_{loc} and if, for every C^1 function ψ with compact support contained in the set $]-\infty,T[\times\mathbb{R}$, it holds that

$$\int_0^T \int_{\mathbb{R}} \{u\cdot\psi_t + f(u)\cdot\psi_x\}dxdt + \int_{\mathbb{R}} u_0(x)\cdot\psi(0,x)dx = 0. \tag{2.4}$$

A weak solution u to (2.3) satisfies

$$u(0,x) = u_0(x) \quad \text{for a.e. } x\in\mathbb{R}.$$

This is a consequence of the fact that u is continuous as a function from $[0,T]$ to L^1_{loc} and of (2.4). We summarize further properties of weak solutions.

Definition 2.3. *A function $u = u(t,x)$ has an approximate jump discontinuity at the point (τ,ξ) if there exist vectors $u^-,u^+\in\mathbb{R}^n$ and $\lambda\in\mathbb{R}$ such that*

$$\lim_{r\to 0^+}\frac{1}{r^2}\int_{-r}^{r}\int_{-r}^{r}\|u(\tau+t,\xi+x)-U(t,x)\|dxdt = 0,$$

where

$$U(t,x) := \begin{cases} u^-, & \text{if } x<\lambda t,\\ u^+, & \text{if } x>\lambda t.\end{cases} \tag{2.5}$$

The function U is called a shock travelling wave.

Theorem 2.4. *Consider a bounded weak solution u to (2.1) with an approximate jump discontinuity at (τ,ξ). Then*

$$\lambda(u^+ - u^-) = f(u^+) - f(u^-). \tag{2.6}$$

A proof can be found in [10].

Equation (2.6), called the Rankine–Hugoniot condition, gives a condition on discontinuities of weak solutions of (2.1) relating the right and left states with the "speed" λ of the "shock." In the scalar case (2.6) is a single equation and, for arbitrary $u^- \neq u^+$, we have

$$\lambda = \frac{f(u^+) - f(u^-)}{u^+ - u^-}.$$

For an $n \times n$ system of conservation laws, (2.6) is a system of n scalar equations.

It is known that weak solutions are in general not unique: let u_0 be the function defined by

$$u_0(x) := \begin{cases} 1, & \text{if } x \geq 0, \\ 0, & \text{if } x < 0. \end{cases}$$

For every $0 < \alpha < 1$, the function $u_\alpha : [0, +\infty[\times \mathbb{R} \to \mathbb{R}$ defined by

$$u_\alpha(t,x) := \begin{cases} 0, & \text{if } x < \frac{\alpha t}{2}, \\ \alpha, & \text{if } \frac{\alpha t}{2} \leq x < \frac{(1+\alpha)t}{2}, \\ 1, & \text{if } x \geq \frac{(1+\alpha)t}{2} \end{cases}$$

is a weak solution to the Burgers equation for all values of α. Therefore the notion of weak solution must be supplemented with admissibility conditions, motivated by physical considerations.

Entropy Admissible Solutions

A first admissibility criterion, coming from physical considerations (see Dafermos [21]), is that of the entropy admissibility condition.

Definition 2.5. *A C^1 function $\eta : \mathbb{R}^n \to \mathbb{R}$ is an entropy for* (2.1) *if it is convex and there exists a C^1 function $q : \mathbb{R}^n \to \mathbb{R}$ such that*

$$D\eta(u) \cdot Df(u) = Dq(u) \tag{2.7}$$

for every $u \in \mathbb{R}^n$. The function q is said to be an entropy flux for η. The pair (η, q) is said to be an entropy–entropy flux pair for (2.1).

Definition 2.6. *A weak solution $u = u(t,x)$ to the Cauchy problem*

$$\begin{cases} u_t + f(u)_x = 0, \\ u(0,x) = u_0(x) \end{cases} \tag{2.8}$$

is said to be entropy admissible if, for every C^1 function $\phi \geq 0$ with compact support in $[0, T[\times \mathbb{R}$ and for every entropy–entropy flux pair (η, q), it holds that

$$\int_0^T \int_{\mathbb{R}} \{\eta(u)\phi_t + q(u)\phi_x\}\, dx\, dt \geq 0. \tag{2.9}$$

Existence of entropies is in general not granted. However, in the scalar case there is a family of convex entropies introduced first by Kruzkov.

Let us consider the scalar Cauchy problem

$$\begin{cases} u_t + f(u)_x = 0, \\ u(0,x) = u_0(x), \end{cases} \tag{2.10}$$

where $f : \mathbb{R} \to \mathbb{R}$ is a C^1 function. In this case the relation between C^1 entropy and entropy flux takes the form

$$\eta'(u) f'(u) = q'(u). \tag{2.11}$$

Therefore if we take a C^1 entropy η, every corresponding entropy flux q has the expression

$$q(u) = \int_{u_0}^{u} \eta'(s) f'(s) ds,$$

where u_0 is an arbitrary element of $\mathbb{R}$.

Definition 2.7. *A weak solution $u = u(t,x)$ to the scalar Cauchy problem* (2.10) *satisfies the Kruzkov entropy admissibility condition if*

$$\int_0^T \int_{\mathbb{R}} \{\|u-k\|\phi_t + sgn\,(u-k)(f(u) - f(k))\phi_x\} dxdt \geq 0$$

for every $k \in \mathbb{R}$ and every C^1 function $\phi \geq 0$ with compact support in $[0,T[\times\mathbb{R}$.

We have the following theorem.

Theorem 2.8. *Let $u = u(t,x)$ be a piecewise C^1 solution to the scalar equation* (2.10). *Then u satisfies the Kruzkov entropy admissible condition if and only if along every line of jump $x = \xi(t)$ the following condition holds. For every $\alpha \in [0,1]$*

$$\begin{cases} f(\alpha u^+ + (1-\alpha)u^-) \geq \alpha f(u^+) + (1-\alpha) f(u^-), & \text{if } u^- < u^+, \\ f(\alpha u^+ + (1-\alpha)u^-) \leq \alpha f(u^+) + (1-\alpha) f(u^-), & \text{if } u^- > u^+, \end{cases} \tag{2.12}$$

where $u^- := u(t,\xi(t)-)$ and $u^+ := u(t,\xi(t)+)$.

For a proof of this theorem see [10]. Equation (2.12) implies that if $u^- < u^+$, then the graph of f remains above the segment connecting $(u^-, f(u^-))$ to $(u^+, f(u^+))$, while if $u^- > u^+$, then the graph of f remains below the segment connecting $(u^-, f(u^-))$ to $(u^+, f(u^+))$ (see Figure 2.2).

Riemann Problem

This section describes the entropy admissible solutions to a Riemann problem, i.e., a Cauchy problem with Heaviside initial data. The construction of solutions to Riemann problems is the first step in a general proof of weak solutions. Let $\Omega \subset \mathbb{R}^n$ be an open set, let $f : \Omega \to \mathbb{R}^n$ be a smooth flux, and consider the system of conservation laws

$$u_t + f(u)_x = 0, \tag{2.13}$$

which we suppose to be strictly hyperbolic.

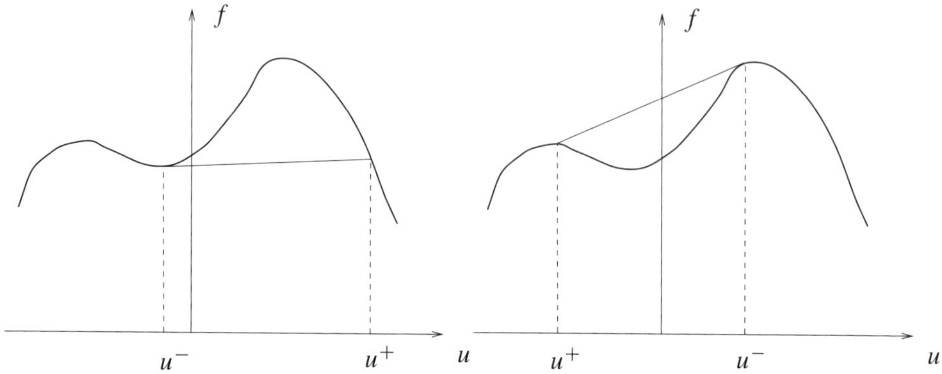

Figure 2.2. *The condition* (2.12) *in the case* $u^- < u^+$ *(left) and the condition* (2.12) *in the case* $u^- > u^+$ *(right).*

Definition 2.9. *A Riemann problem for the system* (2.13) *is the Cauchy problem for the initial datum*

$$u_0(x) := \begin{cases} u^-, & \text{if } x < 0, \\ u^+, & \text{if } x > 0, \end{cases} \tag{2.14}$$

where $u^-, u^+ \in \Omega$.

Remark 2.1.2. *As shown in Section* 2.1, *the solution of Riemann problems is the key step to solving Cauchy problems. In fact to prove existence we use the wave-front tracking method, which, roughly speaking, consists in the following steps:*

1. *approximate the initial condition with piecewise constant solutions;*
2. *at every point of discontinuity solve the corresponding Riemann problem;*
3. *approximate the exact solution to Riemann problems with piecewise constant functions and piece them together to get a function defined until two wave fronts interact together;*
4. *repeat inductively the previous construction starting from the interaction time;*
5. *prove that the functions so constructed converge to a limit function and prove that this limit function is an entropy admissible solution.*

As before we denote by $A(u)$ the Jacobian matrix of the flux f and with $\lambda_1(u) < \cdots < \lambda_n(u)$ the n eigenvalues of the matrix $A(u)$. Let $\{r_1(u), \ldots, r_n(u)\}$, $\{l_1(u), \ldots, l_n(u)\}$ be, respectively, bases of right and left eigenvectors such that the following hold:

1. $\|r_i(u)\| \equiv 1$ for every $u \in \Omega$ and $i \in \{1, \ldots, n\}$;
2. $l_i \cdot r_j = \delta_{ij}$ for every $i, j \in \{1, \ldots, n\}$, where δ_{ij} denotes the Kronecker symbol; that is,

$$\delta_{ij} := \begin{cases} 1, & \text{if } i = j, \\ 0, & \text{if } i \neq j. \end{cases}$$

We introduce the following notation. If $i \in \{1,\ldots,n\}$, then

$$r_i \bullet \lambda_j(u) := \lim_{\varepsilon \to 0} \frac{\lambda_j(u+\varepsilon r_i(u)) - \lambda_j(u)}{\varepsilon},$$

which is the directional derivative of $\lambda_j(u)$ in the direction of $r_i(u)$.

Definition 2.10. *We say that the i-characteristic field ($i \in \{1,\ldots,n\}$) is genuinely nonlinear if*

$$r_i \bullet \lambda_i(u) \neq 0 \qquad \forall u \in \Omega.$$

We say that the i-characteristic field ($i \in \{1,\ldots,n\}$) is linearly degenerate if

$$r_i \bullet \lambda_i(u) = 0 \qquad \forall u \in \Omega.$$

If the ith characteristic field is genuinely nonlinear, then, for simplicity, we assume that $r_i \bullet \lambda_i(u) > 0$ for every $u \in \Omega$.

We consider three cases.

1. Centered rarefaction waves. For $u^- \in \Omega$, $i \in \{1,\ldots,n\}$, and $\sigma > 0$, we denote by $R_i(\sigma)(u^-)$ the solution to

$$\begin{cases} \frac{du}{d\sigma} = r_i(u), \\ u(0) = u^-. \end{cases} \tag{2.15}$$

Let $\bar{\sigma} > 0$. Define $u^+ = R_i(\bar{\sigma})(u^-)$ for some $i \in \{1,\ldots,n\}$. If the ith characteristic field is genuinely nonlinear, then the function

$$u(t,x) := \begin{cases} u^-, & \text{if } x < \lambda_i(u^-)t, \\ R_i(\sigma)(u^-), & \text{if } x = \lambda_i(R_i(\sigma)(u^-))t, \sigma \in [0,\bar{\sigma}], \\ u^+, & \text{if } x > \lambda_i(u^+)t \end{cases} \tag{2.16}$$

is an entropy admissible solution to the Riemann problem

$$\begin{cases} u_t + f(u)_x = 0, \\ u(0,x) = u_0(x), \end{cases}$$

with u_0 defined in (2.14). The function $u(t,x)$ is called a centered rarefaction wave.

Remark 2.1.3. *Notice that, to construct function* (2.16), *$\bar{\sigma}$ must be positive.*

2. Shock waves. Fix $u^- \in \Omega$ and $i \in \{1,\ldots,n\}$. For some $\sigma_0 > 0$, there exist smooth functions $S_i(u_-) = S_i : [-\sigma_0,\sigma_0] \to \Omega$ and $\lambda_i : [-\sigma_0,\sigma_0] \to \mathbb{R}$ such that the following hold:

   (a) $f(S_i(\sigma)) - f(u^-) = \lambda_i(\sigma)(S_i(\sigma) - u^-)$ for every $\sigma \in [-\sigma_0,\sigma_0]$;

   (b) $\|\frac{dS_i}{d\sigma}\| \equiv 1$; $S_i(0) = u^-$, $\lambda_i(0) = \lambda_i(u^-)$; $\frac{dS_i(\sigma)}{d\sigma}|_{\sigma=0} = r_i(u^-)$;

   (c) $\frac{d\lambda_i(\sigma)}{d\sigma}|_{\sigma=0} = \frac{1}{2} r_i \bullet \lambda_i(u^-)$; $\frac{d^2 S_i(\sigma)}{d\sigma^2}|_{\sigma=0} = r_i \bullet r_i(u^-)$.

Let $\bar{\sigma} < 0$. Define $u^+ = S_i(\bar{\sigma})$. If the ith characteristic field is genuinely nonlinear, then the function

$$u(t,x) := \begin{cases} u^-, & \text{if } x < \lambda_i(\bar{\sigma})t, \\ u^+, & \text{if } x > \lambda_i(\bar{\sigma})t \end{cases} \tag{2.17}$$

is an entropy admissible solution to the Riemann problem

$$\begin{cases} u_t + f(u)_x = 0, \\ u(0,x) = u_0(x), \end{cases}$$

with u_0 defined in (2.14). The function $u(t,x)$ is called a shock wave.

Remark 2.1.4. *If we consider $\bar{\sigma} > 0$, then* (2.17) *is again a weak solution, but it does not satisfy the entropy condition.*

3. Contact discontinuities. Fix $u^- \in \Omega$, $i \in \{1,\ldots,n\}$, and $\bar{\sigma} \in [-\sigma_0, \sigma_0]$. Define $u^+ = S_i(\bar{\sigma})$. If the ith characteristic field is linearly degenerate, then the function

$$u(t,x) := \begin{cases} u^-, & \text{if } x < \lambda_i(u^-)t, \\ u^+, & \text{if } x > \lambda_i(u^-)t \end{cases} \tag{2.18}$$

is an entropy admissible solution to the Riemann problem

$$\begin{cases} u_t + f(u)_x = 0, \\ u(0,x) = u_0(x), \end{cases}$$

with u_0 defined in (2.14). The function $u(t,x)$ is called a contact discontinuity.

Remark 2.1.5. *If the ith characteristic field is linearly degenerate, then*

$$\lambda_i(u^-) = \lambda_i(u^+) = \lambda_i(\sigma)$$

for every $\sigma \in [-\sigma_0, \sigma_0]$.

Definition 2.11. *The waves defined in* (2.16), (2.17), *and* (2.18) *are called waves of the ith family.*

For each $\sigma \in \mathbb{R}$ and $i \in \{1,\ldots,n\}$, let us consider the function

$$\psi_i(\sigma)(u_0) := \begin{cases} R_i(\sigma)(u_0), & \text{if } \sigma \geq 0, \\ S_i(\sigma)(u_0), & \text{if } \sigma < 0, \end{cases} \tag{2.19}$$

where $u_0 \in \Omega$. The value σ is called the strength of the wave of the ith family, connecting u_0 to $\psi_i(\sigma)(u_0)$. It follows that $\psi_i(\cdot)(u_0)$ is a smooth function. Moreover, let us consider the composite function

$$\Psi(\sigma_1,\ldots,\sigma_n)(u^-) := \psi_n(\sigma_n) \circ \cdots \circ \psi_1(\sigma_1)(u^-), \tag{2.20}$$

where $u^- \in \Omega$ and $(\sigma_1,\dots,\sigma_n)$ belongs to a neighborhood of 0 in $\mathbb{R}^n$. It is not difficult to calculate the Jacobian matrix of the function Ψ and to prove that it is invertible in a neighborhood of $(0,\dots,0)$. Hence we can apply the implicit function theorem and prove the following result.

Theorem 2.12. *For every compact set $K \subset \Omega$, there exists $\delta > 0$ such that, for every $u^- \in K$ and for every $u^+ \in \Omega$ with $\|u^+ - u^-\| \le \delta$, there exists a unique $(\sigma_1,\dots,\sigma_n)$ in a neighborhood of $0 \in \mathbb{R}^n$ satisfying*

$$\Psi(\sigma_1,\dots,\sigma_n)(u^-) = u^+.$$

Moreover, the Riemann problem connecting u^- with u^+ has an entropy admissible solution, constructed by piecing together the solutions of n Riemann problems.

Functions with Bounded Variation

In this section we give some basic facts about functions with bounded variation. Consider an interval J contained in $\mathbb{R}$ and a function $w : J \to \mathbb{R}$. The total variation of w is defined by

$$T.V.(w) = \sup\left\{\sum_{j=1}^{N} \|w(x_j) - w(x_{j-1})\|\right\}, \tag{2.21}$$

where $N \ge 1$, and the points x_j belong to J for every $j \in \{0,\dots,N\}$ and satisfy $x_0 < x_1 < \cdots < x_N$.

Definition 2.13. *We say that the function $w : J \to \mathbb{R}$ has bounded total variation if $T.V.(w) < +\infty$. We denote with $BV(J)$ the set of all real functions $w : J \to \mathbb{R}$ with bounded total variation.*

Notice that the total variation of a function w is a positive number. If w is a function with bounded total variation, then it is clear that w is a bounded function. The converse is false. In fact every nonconstant periodic and bounded function on $\mathbb{R}$ has total variation equal to $+\infty$. An important property of functions with bounded total variation is the existence of left and right limits for every x of the interior of J.

Lemma 2.14. *Let $w : J \to \mathbb{R}$ be a function with bounded total variation and $\bar{x}$ be a point in the interior of J. Then the limits*

$$\lim_{x\to\bar{x}^-} w(x), \qquad \lim_{x\to\bar{x}^+} w(x)$$

exist. Moreover, the function w has at most countably many points of discontinuity.

The next theorem shows that subsets of $BV(J)$, with uniform bound in total variation, have some compactness properties.

Theorem 2.15 (Helly). *Consider a sequence of functions $w_n : J \to \mathbb{R}^m$. Assume that there exist positive constants C and M such that the following hold:*

1. *T.V.* $(w_n) \leq C$ *for every* $n \in \mathbb{N}$;

2. $\|w_n(x)\| \leq M$ *for every* $n \in \mathbb{N}$ *and* $x \in J$.

Then there exist a function $w : J \to \mathbb{R}^m$ *and a subsequence* w_{n_k} *such that the following hold:*

1. $\lim_{k\to+\infty} w_{n_k}(x) = w(x)$ *for every* $x \in J$;

2. *T.V.* $(w) \leq C$;

3. $\|w(x)\| \leq M$ *for every* $x \in J$.

Theorem 2.16. *Consider a sequence of functions* $w_n : [0,+\infty[\times J \to \mathbb{R}^n$. *Assume that there exist positive constants* C, L, *and* M *such that the following hold:*

1. *T.V.* $(w_n(t,\cdot)) \leq C$ *for every* $n \in \mathbb{N}$ *and* $t \geq 0$;

2. $\|w_n(t,x)\| \leq M$ *for every* $n \in \mathbb{N}$, $x \in J$, *and* $t \geq 0$;

3. $\int_J \|w_n(t,x) - w_n(s,x)\| dx \leq L\|t-s\|$ *for every* $n \in \mathbb{N}$ *and* $t,s \geq 0$.

Then there exist a function $w \in L^1_{loc}([0,+\infty[\times J;\mathbb{R}^n)$ *and a subsequence* w_{n_k} *such that the following hold:*

1. $w_{n_k} \to w$ *in* $L^1_{loc}([0,+\infty[\times J;\mathbb{R}^n)$ *as* $k \to +\infty$;

2. $\int_J \|w(t,x) - w(s,x)\| dx \leq L\|t-s\|$ *for every* $t,s \geq 0$.

Moreover, the values of w *can be uniquely determined by setting*

$$w(t,x) = \lim_{y\to x^+} w(t,y)$$

for every $t \geq 0$ *and* $x \in$ *interior* J. *In this case we have the following:*

1. *T.V.* $(w(t,\cdot)) \leq C$ *for every* $t \geq 0$;

2. $\|w(t,x)\| \leq M$ *for every* $t \geq 0$ *and* $x \in J$.

Wave-Front Tracking and Existence of Solutions

This section deals with the existence of an entropy admissible solution to the Cauchy problem

$$\begin{cases} u_t + [f(u)]_x = 0, \\ u(0,\cdot) = \bar{u}(\cdot), \end{cases} \tag{2.22}$$

where $f : \mathbb{R}^n \to \mathbb{R}^n$ is a smooth flux and $\bar{u} \in L^1(\mathbb{R}^n)$ is bounded in total variation. In order to prove existence, we construct a sequence of approximate solutions using the method called the wave-front tracking algorithm. We start considering the scalar case, while the system case, much more delicate, will only be sketched.

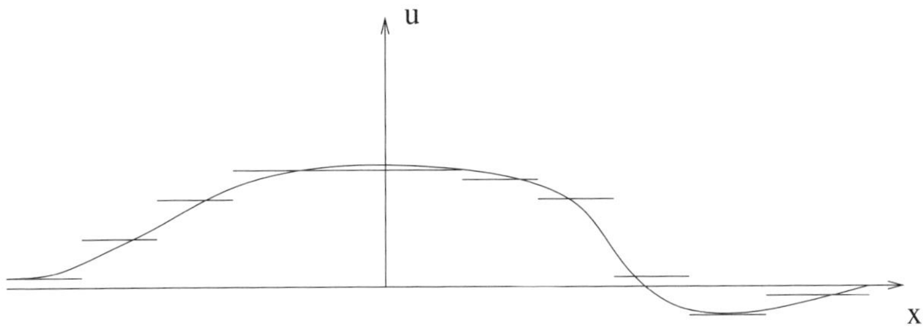

Figure 2.3. *A piecewise constant approximation of the initial datum satisfying* (2.24) *and* (2.25).

The Scalar Case

We assume the following conditions:

(C1) $f : \mathbb{R} \to \mathbb{R}$ is a scalar smooth function;

(C2) the characteristic field is either genuinely nonlinear or linearly degenerate.

It is possible to choose a sequence of piecewise constant functions $(\bar{u}_\nu)_\nu$ such that

$$T.V.(\bar{u}_\nu) \leq T.V.(\bar{u}), \tag{2.23}$$

$$\|\bar{u}_\nu\|_{L^\infty} \leq \|\bar{u}\|_{L^\infty}, \tag{2.24}$$

and

$$\|\bar{u}_\nu - \bar{u}\|_{L^1} < \frac{1}{\nu} \tag{2.25}$$

for every $\nu \in \mathbb{N}$; see Figure 2.3. Fix $\nu \in \mathbb{N}$. By (2.23), $\bar{u}_\nu$ has a finite number of discontinuities, say $x_1 < \cdots < x_N$. For each $i = 1,\dots,N$, we approximately solve the Riemann problem generated by the jump $(\bar{u}_\nu(x_i-), \bar{u}_\nu(x_i+))$ with piecewise constant functions of the type $\phi(\frac{x-x_i}{t})$, where $\phi : \mathbb{R} \to \mathbb{R}$. More precisely, if the Riemann problem generated by $(\bar{u}_\nu(x_i-), \bar{u}_\nu(x_i+))$ admits an exact solution containing just shocks or contact discontinuities, then $\phi(\frac{x-x_i}{t})$ is the exact solution, while if a rarefaction wave appears, then we split it into a centered rarefaction fan, containing a sequence of jumps of size at most $\frac{1}{\nu}$, travelling with a speed between the characteristic speeds of the states connected. In this way, we are able to construct an approximate solution $u_\nu(t,x)$ until a time t_1, where at least two wave fronts interact together; see Figure 2.4.

Remark 2.1.6. *In the scalar case, if the characteristic field is linearly degenerate, then all the waves are contact discontinuities and travel at the same speed. Therefore the previous construction can be done for every positive time.*

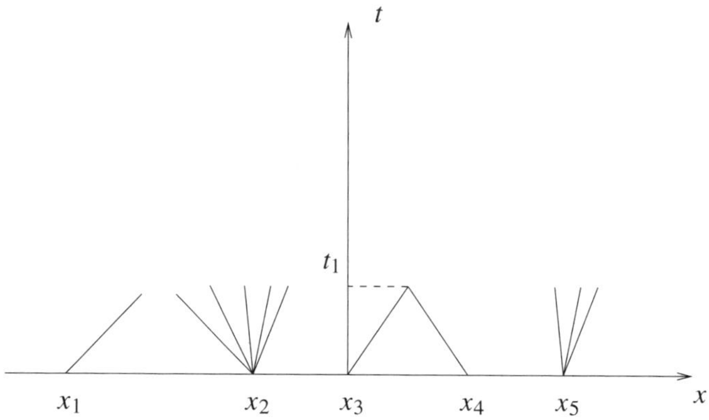

Figure 2.4. *The wave-front tracking construction until the first time of interaction.*

Remark 2.1.7. *Notice that it is possible to avoid that three or more wave fronts interact together at the same time by slightly changing the speed of some wave fronts. This may introduce a small error of the approximate solution with respect to the exact one.*

At time $t = t_1$, $u_\nu(t_1,\cdot)$ is clearly a piecewise constant function. So we can repeat the previous construction until a second interaction time $t = t_2$ and so on. In order to prove that a wave-front tracking approximate solution exists for every $t \in [0,T]$, where T may also be $+\infty$, we need to estimate

1. the number of waves,
2. the number of interactions between waves, and
3. the total variation of the approximate solution.

The first two estimates are concerned with the possibility to construct a piecewise constant approximate solution. The third estimate, instead, is concerned with the convergence of the approximate solutions towards an exact solution. The two first bounds are nontrivial for the vector case, and it is necessary to introduce simplified solutions to Riemann problems and/or nonphysical waves. Finally, we need a compactness argument to obtain the convergence of the sequence of piecewise constant data to any L^1 initial data.

The proofs of the following statements can be found in [10]. The first lemma shows that the number of interactions is finite; the next deals with the total variation followed by two theorems on compactness and existence in the scalar case.

Lemma 2.17. *The number of wave fronts for the approximate solution u_ν is not increasing with respect to the time, and so u_ν is defined for every $t \geq 0$. Moreover, the number of interactions between waves is bounded by the number of wave fronts.*

Lemma 2.18. *The total variation of $u_\nu(t,\cdot)$ is not increasing with respect to the time. Therefore for each $t \geq 0$*

$$T.V.(u_\nu(t,\cdot)) \leq T.V.(\bar{u}). \tag{2.26}$$

Theorem 2.19. *Consider a sequence of functions $u_\nu : [0,\infty[\times\mathbb{R} \to \mathbb{R}$ such that there exist positive constants C, M, and L with the following properties:*

$$T.V.(u_\nu(t,\cdot)) \leq C \quad \text{for every } t \geq 0, \quad \nu \in \mathbb{N}, \tag{2.27}$$

$$\|u_\nu(t,x)\| \leq M \quad \text{for every } t \geq 0, \quad x \in \mathbb{R}, \quad \nu \in \mathbb{N}, \tag{2.28}$$

$$\int_{-\infty}^{\infty} \|u_\nu(t,x) - u_\nu(s,x)\| dx \leq L\|t-s\| \quad \text{for every } t,s \geq 0, \quad \nu \in \mathbb{N}. \tag{2.29}$$

Then there exists a subsequence, converging to some function u in $L^1_{loc}([0,\infty[\times\mathbb{R};\mathbb{R})$, denoted by u_μ. Moreover, we have

$$\int_{-\infty}^{\infty} \|u(t,x) - u(s,x)\| dx \leq L\|t-s\| \quad \text{for every } t,s \geq 0. \tag{2.30}$$

The point values of u can be uniquely determined by requiring that

$$u(t,x) = u(t,x+) := \lim_{y\to x^+} u(t,y) \quad \text{for every } t \geq 0, \quad x \in \mathbb{R}. \tag{2.31}$$

In this case, one has

$$T.V.(u(t,\cdot)) \leq C \quad \text{for every } t \geq 0 \tag{2.32}$$

and

$$\|u(t,x)\| \leq M \quad \text{for every } t \geq 0, \quad x \in \mathbb{R}. \tag{2.33}$$

The main theoretical result on scalar conservation laws is then as follows.

Theorem 2.20. *Let $f : \mathbb{R} \to \mathbb{R}$ be smooth, and let $\bar{u} \in L^1(\mathbb{R})$ with bounded variation. Then there exists an entropy admissible solution $u(t,x)$ to the Cauchy problem* (2.22)*, defined for every $t \geq 0$. Moreover,*

$$\|u(t,\cdot)\|_{L^\infty} \leq \|\bar{u}(\cdot)\|_{L^\infty} \tag{2.34}$$

for every $t \geq 0$.

The case of hyperbolic systems is far more elaborate, and we refer the reader to the references.

Uniqueness and Continuous Dependence

The aim of this section is to illustrate a method to prove uniqueness and Lipschitz continuous dependence by initial data for solutions to the Cauchy problem, controlling for any two approximate solutions u, u' how their distance varies in time. For simplicity we restrict ourselves to the scalar case. The method was introduced in [10] and is based on a Riemannian-type distance on L^1. In [11], the approach was applied to the systems case. Various alternative methods were recently introduced to treat uniqueness, see the bibliographical note, but the one presented here is the more suitable to be used for networks.

The basic idea is to estimate the L^1 distance viewing L^1 as a Riemannian manifold. We consider the subspace of piecewise constant functions and "generalized tangent vectors" consisting of two components (v,ξ), where $v \in L^1$ describes the L^1 infinitesimal displacement, while $\xi \in \mathbb{R}^n$ describes the infinitesimal displacement of discontinuities. For example, take a family of piecewise constant functions $\theta \to u^\theta$, $\theta \in [0,1]$, each of which has the same number of jumps, say, at the points $x_1^\theta < \cdots < x_N^\theta$. Assume that the following functions are well defined (see Figure 2.5),

$$L^1 \ni v^\theta(x) \doteq \lim_{h\to 0} \frac{u^{\theta+h}(x) - u^\theta(x)}{h},$$

and also the numbers

$$\xi_\beta^\theta \doteq \lim_{h\to 0} \frac{x_\beta^{\theta+h} - x_\beta^\theta}{h}, \qquad \beta = 1,\dots,N.$$

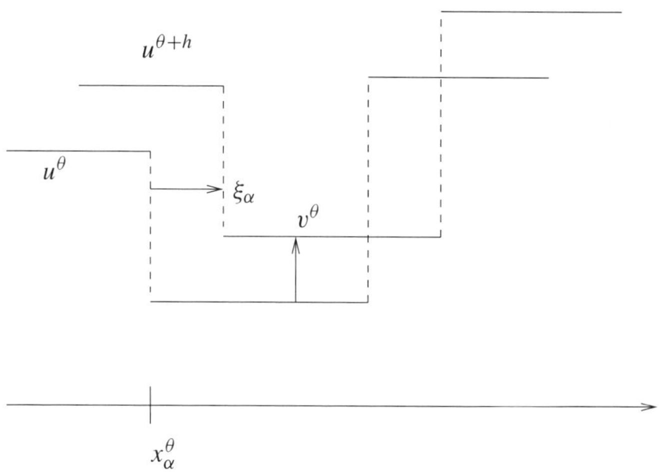

Figure 2.5. *Construction of "generalized tangent vectors."*

Then we say that γ admits tangent vectors $(v^\theta,\xi^\theta) \in T_{u^\theta} \doteq L^1(\mathbb{R};\mathbb{R}^n)\times\mathbb{R}^n$. In general such a path $\theta \to u^\theta$ is not differentiable with respect to the usual differential structure of L^1; in fact if $\xi_\beta^\theta \neq 0$, as $h \to 0$ the ratio $\left[u^{\theta+h}(x) - u^\theta(x)\right]/h$ does not converge to any limit in L^1.

One can compute the L^1 length of the path $\gamma : \theta \to u^\theta$ in the following way:

$$\|\gamma\|_{L^1} = \int_0^1 \|v^\theta\|_{L^1}\, d\theta + \sum_{\beta=1}^N \int_0^1 \left|u^\theta(x_\beta+) - u^\theta(x_\beta-)\right| \left|\xi_\beta^\theta\right| d\theta. \tag{2.35}$$

According to (2.35), in order to compute the L^1 length of a path γ, we integrate the norm of its tangent vector which is defined as follows:

$$\|(v,\xi)\| \doteq \|v\|_{L^1} + \sum_{\beta=1}^N |\Delta u_\beta|\,|\xi_\beta|,$$

where $\Delta u_\beta = u(x_\beta+) - u(x_\beta-)$ is the jump across the discontinuity x_β.

Let us introduce the following definition.

Definition 2.21. *We say that a continuous map* $\gamma : \theta \to u^\theta \doteq \gamma(\theta)$ *from* $[0,1]$ *into* L^1_{loc} *is a regular path if the following holds. All functions* u^θ *are piecewise constant, with the same number of jumps, say, at* $x_1^\theta < \cdots < x_N^\theta$*, and coincide outside some fixed interval* $]-M,M[$*. Moreover,* γ *admits a generalized tangent vector* $D\gamma(\theta) = (v^\theta, \xi^\theta) \in T_{\gamma(\theta)} = L^1(\mathbb{R};\mathbb{R}^n) \times \mathbb{R}^N$*, continuously depending on* θ *.*

Given two piecewise constant functions u and u', call $\Omega(u,u')$ the family of all regular paths $\gamma : [0,1] \to \gamma(t)$ with $\gamma(0) = u$, $\gamma(1) = u'$. The Riemannian distance between u and u' is given by

$$d(u,u') \doteq \inf\left\{\|\gamma\|_{L^1}, \gamma \in \Omega(u,u')\right\}.$$

To define d on all L^1, for given $u,u' \in L^1$ we set

$$\begin{aligned} d(u,u') \doteq \inf \Big\{ &\|\gamma\|_{L^1} + \|u - \tilde{u}\|_{L^1} + \left\|u' - \tilde{u}'\right\|_{L^1} : \\ &\tilde{u}, \tilde{u}' \text{ piecewise constant functions}, \gamma \in \Omega(u,u')\Big\}. \end{aligned}$$

It is easy to check that this distance coincides with the distance of L^1. (For the systems case, one has to introduce weights (see [11]), obtaining an equivalent distance.)

Now we are ready to estimate the L^1 distance among solutions, studying the evolution of norms of tangent vectors along wave-front tracking approximations. Take u,u' as piecewise constant functions and let $\gamma_0(\vartheta) = u^\vartheta$ be a regular path joining $u = u^0$ with $u' = u^1$. Define $u^\vartheta(t,x)$ to be a wave-front tracking approximate solution with initial data u^ϑ and let $\gamma_t(\vartheta) = u^\vartheta(t,\cdot)$.

One can easily check that, for every γ_0 (regular path) and every $t \geq 0$, γ_t is a regular path. If we can prove

$$\|\gamma_t\|_{L^1} \leq \|\gamma_0\|_{L^1}, \tag{2.36}$$

then for every $t \geq 0$

$$\left\|u(t,\cdot) - u'(t,\cdot)\right\|_{L^1} \leq \inf_{\gamma_t} \|\gamma_t\|_{L^1} \leq \inf_{\gamma_0} \|\gamma_0\|_{L^1} = \left\|u(0,\cdot) - u'(0,\cdot)\right\|_{L^1}. \tag{2.37}$$

To obtain (2.36), and hence (2.37), it is enough to prove that, for every tangent vector $(v,\xi)(t)$ to any regular path γ_t, one has

$$\|(v,\xi)(t)\| \leq \|(v,\xi)(0)\|; \tag{2.38}$$

i.e., the norm of a tangent vector does not increase in time. Moreover, if (2.37) is established, then uniqueness and Lipschitz continuous dependence of solutions to Cauchy problems are straightforwardly achieved passing to the limit on the wave-front tracking approximate solutions.

Let us now estimate the increase of the norm of a tangent vector. In order to achieve (2.38), we fix a time $\bar{t}$ and treat the following cases:

Case 1. No interaction of waves takes place at $\bar{t}$.

Case 2. Two waves interact at $\bar{t}$.

In Case 1, denote by x_β, σ_β, and ξ_β, respectively, the positions, sizes, and shifts of the discontinuities of a wave-front tracking approximate solution. Following [11] we get

$$\frac{d}{dt}\left\{\int |v(t,x)|\,dx + \sum_\beta |\xi_\beta|\,|\sigma_\beta|\right\}$$

$$= -\left\{\sum_\beta \left(\lambda(\rho^-) - \dot{x}_\beta\right)|v^-| + \sum_\beta \left(\dot{x}_\beta - \lambda(\rho^+)\right)|v^+|\right\}$$

$$+ \sum_\beta D\lambda(\rho^-,\rho^+)\cdot(v^-,v^+)\left(\mathrm{sign}\xi_\beta\right)|\sigma_\beta|,$$

with $\sigma_\beta = \rho^+ - \rho^-$, $\rho^\pm \doteq \rho(x_\beta\pm)$, and similarly for $v^\pm$. If the waves respect the Rankine–Hugoniot conditions, then

$$D\lambda(\rho^-,\rho^+)(v^-,v^+) = \left(\lambda(\rho^-) - \dot{x}_\beta\right)\frac{v^-}{|\sigma_\beta|} + \left(\dot{x}_\beta - \lambda(\rho^+)\right)\frac{v^+}{|\sigma_\beta|}$$

and

$$\frac{d}{dt}\left\{\int |v(t,x)|\,dx + \sum_\beta |\xi_\beta|\,|\sigma_\beta|\right\} \leq 0. \tag{2.39}$$

In the wave-front tracking algorithm the Rankine–Hugoniot condition may be violated for rarefaction fans. However, this results in an increase of the distance which is controlled in terms of $1/\nu$ (the size of a rarefaction shock) and tends to zero with $\nu \to \infty$.

Let us now pass to Case 2. First, we have the following.

Lemma 2.22. *Consider two waves, with speeds λ_1 and λ_2, respectively, that interact together at $\bar{t}$ producing a wave with speed λ_3. If the first wave is shifted by ξ_1 and the second wave by ξ_2, then the shift of the resulting wave is given by*

$$\xi_3 = \frac{\lambda_3 - \lambda_2}{\lambda_1 - \lambda_2}\xi_1 + \frac{\lambda_1 - \lambda_3}{\lambda_1 - \lambda_2}\xi_2. \tag{2.40}$$

Moreover, we have that

$$\Delta\rho_3\,\xi_3 = \Delta\rho_1\,\xi_1 + \Delta\rho_2\,\xi_2, \tag{2.41}$$

where $\Delta\rho_i$ are the signed strengths of the corresponding waves.

From (2.41) it follows that

$$|\Delta\rho_3\xi_3| \leq |\Delta\rho_1|\,|\xi_1| + |\Delta\rho_2|\,|\xi_2|,$$

from which

$$\left\|(v,\xi)(\bar{t}+)\right\| \leq \left\|(v,\xi)(\bar{t}-)\right\|. \tag{2.42}$$

2.2 Numerical Discretization of Conservation Laws

In order to obtain numerical results for the models based on partial and ordinary differential equations they have to be discretized first. Based on the actual models treated we discuss a few numerical schemes which can be used as discretization. The following discussion just gives a glimpse of the existing schemes and theory, and we refer the reader to standard textbooks such as [63] for further references and more details. In fact, we limit the presentation to the simplest first-order schemes in opposition to existing and well-understood higher-order schemes. The presented schemes below will be useful in particular for optimization purposes since they yield linear equations.

The arising supply chain models are given, for example, by (5.1), (5.3), and (5.51). We apply first-order finite-difference methods to obtain a numerical approximation. We discuss the scheme in the general situation of

$$U_t + AU_x = 0 \tag{2.43}$$

and

$$U_t + F(U)_x = 0, \tag{2.44}$$

where U can be a vector-valued function $U : (t,x) \to \mathbb{R}^n$ and $A \in \mathbb{R}^{n\times n}$. If A and F are real valued, then the equation is called a scalar equation. The model (5.3) is of the type (2.43), and the model (5.51) and (5.1) fits into the context of (2.44). At first, we discuss numerical schemes for (2.43). We discretize the t-x-plane by discrete mesh points (t_j, x_i) with $t_j = j\ \Delta t$ and $x_i = i\ \Delta x$, where Δt and Δx are the mesh widths. The widths are assumed to be constant to simplify the presentation. The numerical approximation of the solution of (2.43) is denoted by $U_i^j = U(t_j, x_i)$. The initial data $U(0,x) = U_0(x)$ is discretized using $U_i^0 = U_0(x_i)$. From this initial data we use a time-marching procedure to construct the approximation U_i^j with $j \geq 1$. For (2.43) we use the following first-order upwind scheme to obtain U_i^{j+1} from U_i^j. In the case of $A \in \mathbb{R}$ and $A \geq 0$ we set

$$U_i^{j+1} = U_i^j - \frac{\Delta t}{\Delta x} A \left(U_i^j - U_{i-1}^j \right). \tag{2.45}$$

This discretization is motivated by the following calculations. The exact solution of (2.43) with initial data $u(0,x) = u_0(x)$ in the case of a scalar conservation law is given by

$$u(t,x) = u_0(x - At). \tag{2.46}$$

Hence, the initial profile u_0 is transported with velocity A. The sign of A describes in which direction the information propagates. In our examples V^e plays the role of A and is hence positive. In order to capture the flow of information for $A \geq 0$ we use in (2.45) one-sided differences to approximate u_x. It can be proved (see, e.g., [63]) that the presented discretization is a first-order in Δt and Δx convergent scheme for (2.43). The convergence of the discrete approximation U_i^j to the solution of (2.43) is proved under a stability condition known as the Courant–Friedrichs–Lewy (CFL) condition: the CFL condition for the scheme (2.45) reads

$$\|A\| \frac{\hat{\mathrm{E}}\Delta t}{\Delta x} \leq 1. \tag{2.47}$$

The condition can be explained by discussing the domain of dependence of solutions u to (2.43); cf. Figure 7.4. Due to (2.46) the solution at (t,x) depends only on the point $x - At$. Hence, it is reasonable to require that the domain of dependence of the discrete approximation (2.45) cover the domain of dependence of the continuous solution. For general three-point schemes such as (2.45) the CFL condition guarantees that the characteristics of (2.43) are within the computational domain of dependence. The CFL condition is in general only necessary but not sufficient for stability; see, e.g., [63, Chapter 10]. Equation (5.3) is discretized using this scheme for any nonnegative V^e. Furthermore, (5.1) is discretized by (2.45) since the flux function is monotone increasing. However, the model (5.51) cannot be discretized by (2.45). The nonlinearity in the flux function of (5.51) might give rise to information propagating in two directions, opposed to the case of (2.43) where in the scalar case the sign of A determines the direction of propagation of information. For these general situations we use the first-order Lax–Friedrichs discretization which is for (2.44) given by

$$U_i^{j+1} = \frac{1}{2}\left(U_{i-1}^j + U_{i+1}^j\right) - \frac{\Delta t}{2\Delta x}\left(F(U_{i+1}^j) - F(U_{i-1}^j)\right). \tag{2.48}$$

The scheme applies central differences for $F(U)_x$. The Lax–Friedrichs scheme is convergent under the CFL condition. It is first-order convergent in Δt and Δx. It can also be used for (2.43); however, the numerical resolution of (2.48) is three times more diffusive. The diffusion roughly describes the ability of the scheme to reproduce discontinuous solutions: the less diffusion the scheme enjoys, the sharper the resolution of the discontinuity. However, there is no need to know the wave speeds of (2.44) when using (2.48). As stated above these two numerical schemes present only the basic numerical schemes for computing approximate solutions. Additional information concerning convergence properties, higher-order schemes, and finite-volume methods is found in [36, 63] and the references therein. Another possible discretization is based on the Godunov scheme. We present the scheme for scalar conservation laws, i.e., $F : \mathbb{R} \to \mathbb{R}$. The Godunov scheme is based on the construction of the Riemann problem for (6.6). The Riemann problem, $[U_L, U_R]$, is the initial value problem for initial data given by a jump discontinuity

$$U(0,x) = \begin{cases} U_L, & x < 0, \\ U_R, & x > 0, \end{cases} \tag{2.49}$$

and it has a unique entropy solution

$$U(t,x) = U_R\left(\frac{x}{t}; U_L, U_R\right). \tag{2.50}$$

We discretize $[0,\infty) \times \mathbb{R}$ by a time mesh length Δt and a spatial mesh length Δx, and we let $t_n = n\Delta t$ and $x^j = j\Delta x$, so that (t_n, x^j) denotes the mesh points of the approximate solution $v_\Delta(t,x) = v_n^j$. Starting by the approximation $v_n = (v_n^j)_{j\in\mathbb{Z}}$ of $U(t_n,\cdot)$, with v a column vector of $\mathbb{R}^2$, an approximation v_{n+1}^j, with $j \in \mathbb{Z}$, of $U(t_{n+1},\cdot)$ can be defined as follows:

- extension of the sequence v_n as a piecewise constant function $v_\Delta(t_n,\cdot)$:

$$v_\Delta(t_n,\cdot) = v_n^j, \quad x^{j-1/2} < x < x^{j+1/2}; \tag{2.51}$$

- solution of the Cauchy problem

$$\begin{cases} U_t + F(U)_x = 0, \ x \in \mathbb{R}, \ t > 0, \\ U(0,x) = v_\Delta(t_n, \cdot) \end{cases} \tag{2.52}$$

in the cell $(t_n, t_{n+1}) \times (x^{j-1}, x^j)$;

- computation of the solution as the average value of the preceding solution in the interval $(x^{j-1/2}, x^{j+1/2})$ obtained projecting $U(\Delta t, \cdot)$ onto the piecewise constant functions:

$$v_{n+1}^j = \frac{1}{\Delta x} \int_{x^{j-1/2}}^{x^{j+1/2}} U(\Delta t, x) dx. \tag{2.53}$$

To avoid the interaction of waves in two neighboring cells before time Δt, we impose a CFL condition like

$$\frac{\Delta t}{\Delta x} \max\{|\lambda_0|, |\lambda_1|\} \leq \frac{1}{2}, \tag{2.54}$$

where λ_0 and λ_1 are the eigenvalues. Since in this case the eigenvalues are such that $|\lambda_0| = 1, |\lambda_1| \leq 1$, the CFL condition reads as

$$\frac{\Delta t}{\Delta x} \leq \frac{1}{2}. \tag{2.55}$$

The solution of (2.52) is obtained by solving a sequence of neighboring Riemann problems, and we have

$$U(t,x) = U_R\left(\frac{x - x^{j+1/2}}{\Delta t}; v_n^j, v_n^{j+1}\right), \quad x^j < x < x^{j+1}, \quad j \in \mathbb{Z}. \tag{2.56}$$

Then, integrating (2.52) over the rectangle $(0, \Delta t) \times (x^{j-1/2}, x^{j+1/2})$ we can obtain a more explicit expression of the scheme. Since the function is piecewise smooth, we get

$$\int_{x^{j-1/2}}^{x^{j+1/2}} (U(\Delta t, 0) - U(0,x)) dx + \int_0^{\Delta t} (F(U(t, x^{j+1/2} - 0)) - F(U(t, x^{j-1/2} + 0))) dt = 0. \tag{2.57}$$

Now, using (2.51) and projecting the solution on piecewise constant functions we obtain

$$\Delta x (v_{n+1}^j - v_n^j) + \int_0^{\Delta t} (F(U(t, x^{j+1/2} - 0)) - F(U(t, x^{j-1/2} + 0))) dt = 0, \tag{2.58}$$

and, recalling (2.56), we derive

$$v_{n+1}^j = v_n^j - \frac{\Delta t}{\Delta x} \{F(U_R(0-; v_n^j, v_n^{j+1})) - F(U_R(0+; v_n^{j-1}, v_n^j))\}. \tag{2.59}$$

Since the function $\xi \to F(U_R(\xi; U_L, U_R))$ is continuous at the origin due to the Rankine–Hugoniot conditions (see [35]), the Godunov scheme can be written in the form

$$v_{n+1}^j = v_n^j - \frac{\Delta t}{\Delta x} \{F(U_R(0; v_n^j, v_n^{j+1})) - F(U_R(0; v_n^{j-1}, v_n^j))\}, \tag{2.60}$$

and the numerical flux computed in $V = (v_1, v_2)$ and $W = (w_1, w_2)$ is

$$G(V,W) = F(U_R(0;V,W)). \tag{2.61}$$

The numerical flux can be written in a general form as

$$G(V,W) = \begin{cases} \min_{z_1 \in [v_1, w_1]} F(Z), & \text{if } v_1 \le w_1, \\ \\ \max_{z_1 \in [w_1, v_1]} F(Z), & \text{if } w_1 \le v_1, \end{cases}$$

where the second variable z_2 in $Z = (z_1, z_2)$ is assumed to be fixed. The final expression of the Godunov scheme for the problem (2.52) is

$$v_{n+1}^j = v_n^j - \frac{\Delta t}{\Delta x}\big(G(v_n^j, v_n^{j+1}) - G(v_n^{j-1}, v_n^j)\big). \tag{2.62}$$

More precisely, for the system (6.3), the scheme reads as

$$\begin{cases} \rho_{n+1}^j = \rho_n^j - \frac{\Delta t}{\Delta x}(g(\rho_n^j, \rho_n^{j+1}) - g(\rho_n^{j-1}, \rho_n^j)), \\ \mu_{n+1}^j = \mu_b^j + \frac{\Delta t}{\Delta x}(\mu_n^{j+1} - \mu_n^j), \end{cases} \tag{2.63}$$

where we indicate the approximate values of $\rho(t,x)$ and $\mu(t,x)$ on the numerical grid as, respectively, ρ_n^j and μ_n^j for $j = 0, \dots, L$ and $n = 0, \dots, M-1$. Notice that the Godunov scheme for the second equation reduces to the forward upwind scheme.

Chapter 3

Basic Queueing Models

Manufacturing systems can be described by using a number of mathematical approaches. All of these approaches present a powerful tool to study and analyze the behavior of production systems under specific conditions. Basically, we distinguish between stationary, also called steady state, and instationary models which are time dependent.

A well-known class of stationary models are queuing theory models; see [9, 16]. These models permit the derivation and calculation of several performance measures, including the mean waiting time of parts in the system, the proportion of time the processors are busy, and the probability of encountering the system in certain states.

In contrast, instationary models predict the time evolution of parts and include a dynamics inside the different production steps. These dynamic models can again be divided into two classes: discrete event and fluid models. Discrete event models [2, 7] are an evident access to model production networks. Since these models are based on considerations of individual parts, the principal drawback of these models, however, is their enormous computational effort. A cost-effective alternative to discrete event models is fluid-like network models using partial differential equations describing averaged quantities like density and average velocity. They have been originally developed and extensively analyzed in the field of traffic flow; see, for instance, [17, 33, 40, 47, 53]. In recent years, the ideas of traffic flow models have been extended to supply chains, especially to production networks. We refer the reader to [1, 2, 3, 4, 23, 26] for an overview.

Throughout this book we will focus on continuous equations of nonlinear type which, on the one hand, provide a basis for the inclusion of nonlinear dynamics and, on the other hand, can be used for simulation as well as optimization purposes. All models under investigation are purely deterministic and do not involve any stochastic phenomena.

In the following chapter we present the derivation of a basic continuous supply chain model which enables describing time-dependent production processes. This introductory model was presented by Armbruster, Degond, and Ringhofer [2] in 2006. Therein, they started their investigations from a discrete event simulation and derived in the continuum limit a conservation law for the part density of the form $\partial_t \rho + \partial_x \min\{V\rho, \mu\}$ where the flux function f is cut off by the maximal capacity of suppliers. We will see that possible bottlenecks are treated as unlovely delta distributions in the part density and therefore restrict the analytical and numerical opportunities of the model. How to avoid this discomfort in general will be part of Chapter 5.

This chapter is organized as follows. Section 3.2 introduces basic terminology and the way of working with discrete event simulations. In Section 3.3 we focus on the derivation of a continuous supply chain model, namely a conservation law for the part density, and point out its fundamental properties. All models are illustrated and explained using small examples. For further details and especially the proofs we refer the reader to the original work of Armbruster, Degond, and Ringhofer [2].

3.1 Motivation

A very effective and common approach for the mathematical modeling of supply chains is so-called discrete event simulations. They provide a basis for an accurate description of the underlying process. The main idea of these models is to track goods from origins (suppliers) to destinations (customers). From a mathematical point of view to track means to compute arrival times of each good over the whole network. Clearly, these times are dependent on the current customer demand and the order policies used by the suppliers. Let us consider the situation of a serial supply chain (see Figure 3.1) consisting of suppliers $S_1, \ldots, S_M$ where each supplier has user-defined properties such as processing time T (measured in time units) and maximal processing rate μ (measured in goods per time unit).

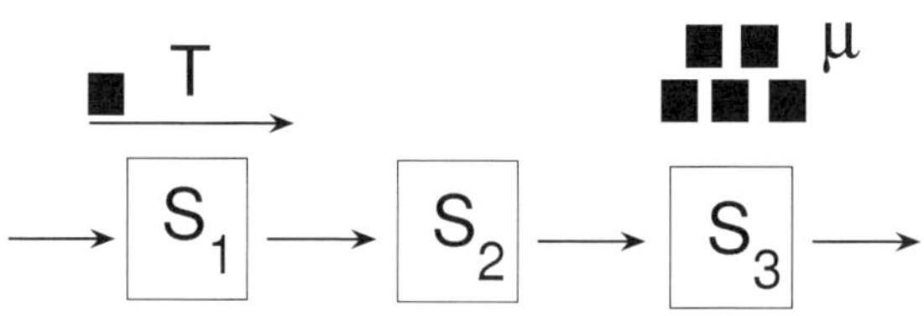

Figure 3.1. *Individual processing parameters.*

Naturally, suppliers may have different processing rates, so it may happen that goods have to wait until the next operations can be performed. The movement of goods is now determined by calculating the arrival times at all suppliers according to a fundamental time recursion. Curves of cumulative count help to analyze how many parts are currently in process. Although discrete event simulation models reflect the most accurate way of modeling a supply network, the computational effort is extremely high in particular for a large number of suppliers and goods. One way out of the dilemma is models based on partial differential equations. They are used to describe the propagation of part flow (parts per time unit), which means that parts are no longer considered individually. Following only a few major guidelines we derive formally a continuous model from the aforementioned discrete event model. This new model is computationally feasible and describes accurately the dynamic behavior of the system. As we will see, this model serves as a starting point for extensions to more general supply chains.

3.2 Discrete Event Simulation

Discrete event simulation depicts one way of building up models to observe the dynamic behavior of manufacturing systems. We begin with introducing some elementary principles of discrete event simulation models by considering the time evolution of individual parts. Such a production system is organized as an assembly line consisting of processors and queues (see Figure 3.1) where each processor is responsible for a certain production step and parts are moved between processors. Our modeling approach is based on the following assumptions:

(A1) No parts are lost or generated inside a single processor.

(A2) Each processor has a maximal capacity measured in units of parts.

(A3) A queue in front of each processor is used to store goods (see Figure 3.2).

We consider a production line with M processors (or suppliers) where a single processor m is linked only with the next processor $m+1$; cf. Figure 3.3. Basically, every processor is characterized by its processing time $T(m)$ as well as its maximal capacity (production rate) $\mu(m)$. The processing time $T(m)$ is the time which is needed to produce a single part or, respectively, to finish one production step. The maximum capacity $\mu(m)$ is defined as the maximal number of parts per time unit which can be handled by one processor. According to definitions, the product $T(m) \cdot \mu(m)$ states the absolute number of parts a processor is able to deal with. Note that the simplest choice would be to set $\mu(m) = \frac{1}{T(m)}$; i.e., the processor handles only one part per time unit.

In this model, both parameters are fixed and deterministic, although random breakdowns of processors, for instance maintenance and repair, would be possible. The unbounded queue of the processor temporarily buffers all parts going into this processor. More precisely, in the case of an empty queue, parts are given instantaneously into the processor, and otherwise, if there are any parts in the queue and the processor already works on its maximal capacity, the newly arriving parts are buffered and wait to be processed.

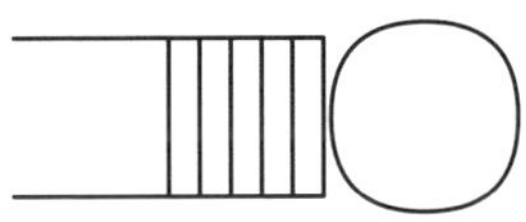

Figure 3.2. *Queue and processor.*

The dynamics of discrete event simulations are given by tracking parts through the network and calculating arrival times at processors. The total number of parts in the system is denoted by N. We define the arrival time of one part n at the beginning of processor m as a_n^m. For a precise description of the discrete event algorithm, we need to determine additional time measures such as release and leaving times. The release time b_n^m is the time that a part n is fed into the processor m. Furthermore, the leaving time e_n^m is the time that a part n leaves processor m and arrives at processor $m+1$; see Figure 3.3.

Obviously, the computation of release times b_n^m is dependent on the actual queue length. By assuming the FIFO (first in first out) policy, two cases of queue states can be distinguished: Either the queue is empty or it is nonempty. If the queue is empty, part n is immediately passed into processor m and, consequently, the release time corresponds to the arrival time, i.e., $b_n^m = a_n^m$. This is the first argument of the max function in (3.1). Otherwise, if the queue is nonempty, part n has to wait and the waiting time is determined by the inverse of the processing rate. This statement can be found in the second argument

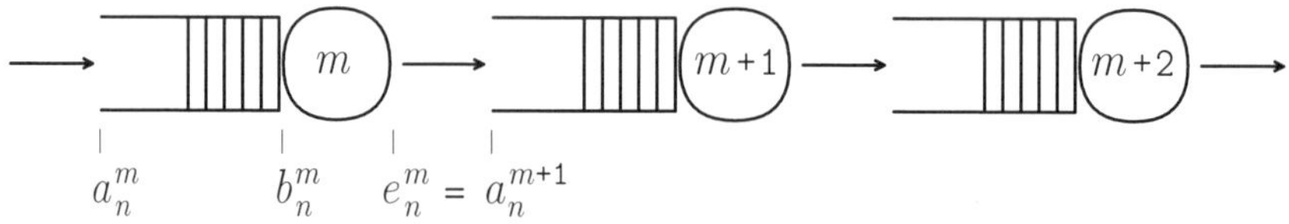

Figure 3.3. *A serial production line.*

of the max function in (3.1):

$$b_n^m = \max\left\{a_n^m, b_{n-1}^m + \frac{1}{\mu(m)}\right\}. \tag{3.1}$$

To determine the leaving times e_n^m, we have to use the information that a certain production step needs the processing time $T(m)$. More precisely, after releasing one part into the processor, it takes the time $T(m)$ to produce it. Hence, we calculate the exit time by

$$e_n^m = b_n^m + T(m). \tag{3.2}$$

Combining the results (3.1) and (3.2) we yield

$$e_n^m = \max\left\{a_n^m + T(m), e_{n-1}^m + \frac{1}{\mu(m)}\right\}, \quad m = 0,1,\dots, \quad n \geq 1. \tag{3.3}$$

In the next step, we change the notation in order to develop a time recursion for the arrival times only. Then, the variable $\tau(n,m)$ denotes the arrival time of part n at processor m, i.e., $a_n^m \to \tau(n,m)$. Furthermore, the exit time e_n^m changes to $e_n^m \to \tau(n,m+1)$, in analogy to the assumption that leaving processor m corresponds to entering processor $m+1$. Based on (3.3) we obtain

$$\tau(n,m+1) = \max\left\{\tau(n,m) + T(m),\ \tau(n-1,m+1) + \frac{1}{\mu(m)}\right\}, \tag{3.4}$$
$$n \geq 1, \quad m = 0,\dots,M-1.$$

To work with a well-defined recursion we have to include initial and boundary conditions of types $\tau(n,0) = \tau^A(n)$, $n \geq 0$, and $\tau(0,m) = \tau^I(m)$, $m = 0,\dots,M$, which can be specified as follows:

$$f^{in}(\tau^A(n)) = \frac{1}{\tau^A(n+1) - \tau^A(n)}, \tag{3.5a}$$

$$\tau^I(m+1) = \tau^I(m) + T(m). \tag{3.5b}$$

The fixed inflow profile f^{in}, prescribed at the first processor $m = 0$, is the rate at which parts enter the system. Then, rewriting the first condition, (3.5a) describes the time when part n arrives at the first processor. By assuming an empty network in the beginning, (3.5b) provides the time when the first part $n = 0$ arrives at processor $m+1$ starting at processor m. Summarizing, the discrete event simulation is completely described with (3.4), (3.5a), and (3.5b).

Remark 3.2.1. *The waiting time $Q(n,m)$ of part n in the queue of processor m is defined as*

$$Q(n,m) = b_n^m - a_n^m \overset{(3.2)}{=} e_n^m - T(m) - a_n^m = \tau(n,m+1) - T(m) - \tau(n,m). \tag{3.6}$$

In the case of empty queues, according to (3.1)*, there will be no waiting time, i.e.,* $Q(n,m) = 0$.

Using the concept of Newell-curves (see [23, 67]), the time recursion (3.4) can be evaluated. Newell-curves (shortly N-curves) describe how the information of (3.4) can be organized to facilitate the computation of performance measures, e.g., the work in progress (WIP). More detailed, N-curves $U(t,m)$ are curves of cumulative counts using the Heaviside function:

$$H(y) = \begin{cases} 0, & \text{if } y < 0, \\ 1, & \text{if } y \geq 0. \end{cases}$$

In the context of supply chains the evaluation of N-curves is given by

$$U(t,m) = \sum_{n=0}^{\infty} H(t - \tau(n,m)), \quad t > 0, \quad m = 0,\dots,M. \tag{3.7}$$

In other words, the N-curve $U(t,m)$ yields the number of parts passing from processor $m-1$ to processor m up to any time t. The difference of two consecutive N-curves yields the number of parts actually produced in processor m (including all parts in the queue as well). This measure is also known as the WIP and is denoted by $W(t,m)$ (see Figure 3.4):

$$W(t,m) = U(t,m) - U(t,m+1), \quad m = 0,\dots,M-1. \tag{3.8}$$

We will see that the WIP can be associated with the part density $\rho(t,x)$ later on; i.e., $W = \int \rho\, dx$ holds.

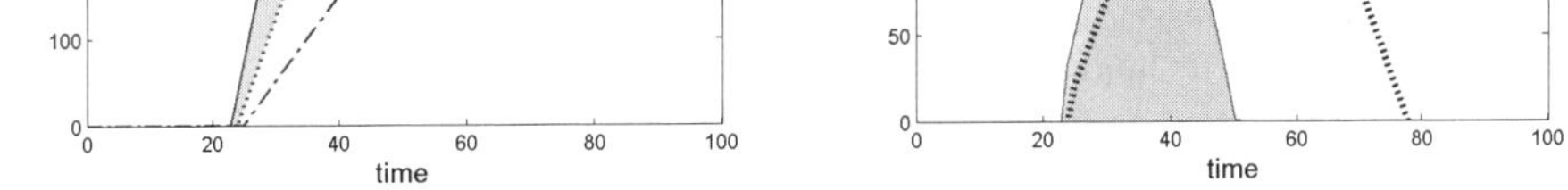

Figure 3.4. *Numerical example of two consecutive processors. The N-curves (left picture) are measured at three significant points: entry S_1, exit S_1 (entry S_2, respectively), and exit S_2. In each case, the area between two N-curves is marked, and the difference, the actual WIP $W(t,m)$, is shown in the right picture.*

If we now consider the derivative of $W(t,m)$ with respect to t, we get

$$\frac{d}{dt}W(t,m) = \frac{d}{dt}U(t,m) - \frac{d}{dt}U(t,m+1) \tag{3.9a}$$

$$= \sum_{n=0}^{\infty}\delta(t-\tau(n,m)) - \sum_{n=0}^{\infty}\delta(t-\tau(n,m+1)) \tag{3.9b}$$

$$= F(t,m) - F(t,m+1). \tag{3.9c}$$

By definition, the flux $F(t,m)$ from supplier S_{m-1} to S_m is given by the time derivative of $U(t,m)$. Since $U(t,m)$ involves generically the Heaviside function, the flux can be interpreted as a superposition of delta distributions. To avoid this inconvenience, the conservation law (3.9) will be replaced by continuous averaged quantities in the following, where the dependence on individual parts is completely removed.

Concluding the discrete event simulation model, some remarks are in order.

1. The functions $U(t,m)$, $F(t,m)$, and $W(t,m)$ will always be nonnegative.

2. By introducing additional N-curves, the WIP can be split into the number of parts in the queue and the number of parts actually produced inside the processor.

3. The concept of computing arrival and leaving times can be easily adapted to more complex networks.

4. It is extremely expensive to compute discrete event simulations for a large number of parts, i.e., $N \to \infty$.

The latter is prevented by the derivation of a conservation law based on the time recursion (3.4). This means exactly the computation of continuous quantities where the WIP will be replaced by a function called part density.

3.3 Capacity-Restricted Conservation Law

To start introducing concepts of continuous supply chain models, we briefly summarize the famous work of Armbruster, Degond, and Ringhofer [2]. Therein, the authors propose a partial differential equation (a conservation law for the part density) for interacting processors in a serial production line. In Chapter 5 this fundamental model provides the basis for ongoing investigations—a general supply network model.

Towards a continuous formulation under the assumption of conservation of parts the following steps are performed: A constitutive relation between the density ρ and the flux f is directly derived from and with recursion (3.4). By a reformulation of the problem, which is necessary to prevent analytical difficulties, it can be shown that the asymptotic limit leads in fact to a partial differential equation. The rather artificial assumption of a huge number of processors is enabled by additionally introducing virtual processors.

Let us begin. First, each processor is mapped onto a gridpoint in space x_m such that approximate expressions for the density ρ and flux f can be derived from the arrival times $\tau(n,m)$ in (3.4). How is this done? Let us go back to the definition of the fluxes $F(t,x_m)$

in (3.9). Formally, we multiply the flux by a smooth test function $\psi(t)$ and consider the integral

$$\int_{\tau^I(m)}^{\infty} \psi(t)F(t,x_m)\,dt = \sum_{n=0}^{\infty} \int_{\tau^I(m)}^{\infty} \psi(t)\delta(t-\tau(n,m)) = \sum_{n=0}^{\infty} \psi(\tau(n,m)). \tag{3.10}$$

It is even possible to rewrite the same integral as a Riemann sum

$$\int_{\tau^I(m)}^{\infty} \psi(t)F(t,x_m)\,dt = \sum_{n=0}^{\infty} \psi(\tau(n,m))\Delta_n\tau(n,m) f(\tau(n,m),x_m), \tag{3.11}$$

where the increment is given by $\Delta_n\tau(n,m) = \tau(n+1,m) - \tau(n,m)$ and further, as a consequence of (3.10), the function $f(\tau(n,m),x_m)$ is determined to be the inverse of $\Delta_n\tau(n,m)$. Since in the limit $\Delta_n\tau \to 0$ the approximation

$$\int_{\tau^I(m)}^{\infty} \psi(t)F(t,x_m)\,dt \approx \int_{\tau^I(m)}^{\infty} \psi(t)f(t,x_m)\,dt$$

is valid, we call the function f the approximate flux for $t = \tau(n,m)$:

$$f(\tau(n,m),x_m) = \frac{1}{\tau(n+1,m)-\tau(n,m)}, \qquad n \geq 0, \quad m = 0,\ldots,M. \tag{3.12}$$

Assuming now that the arrival times τ are continuously distributed, i.e., τ is given in terms of continuous variables, we replace the original term by $\tau(y,x)$ and rewrite the above expression for the approximate flux as $f(\tau(y,x),x) = \frac{1}{\partial_y\tau(y,x)}$. In a similar way, we try to identify the approximate density ρ. An analytical computation shows that the assumption of mass conservation is fulfilled only for a special choice of the part density ρ:

$$\frac{d}{dx} f(\tau(y,x),x) = \frac{\partial}{\partial t} f(\tau(y,x),x)\frac{\partial}{\partial x}\tau(y,x) + \frac{\partial}{\partial x} f(\tau(y,x),x) \stackrel{!}{=} 0.$$

This looks similar to the common notation of a conservation law if the derivative with respect to the first argument is denoted by t. Associating exactly this term to be the approximate density ρ we can deduce that the latter is uniquely defined as $\rho = \frac{\partial_x\tau}{\partial_y\tau}$. Thus, on a discrete level, the approximate density is then

$$\rho(\tau(n,m+1),x_m) = \frac{\tau(n+1,m+1)-\tau(n+1,m)}{h_m(\tau(n+1,m+1)-\tau(n,m+1))}, \tag{3.13a}$$
$$n \geq 0, \quad m = 0,\ldots,M-1,$$

$$h_m := x_{m+1} - x_m, \quad m = 0,\ldots,M-1. \tag{3.13b}$$

At this level, the switching between continuous and discrete variables and functions, respectively, is necessary to formally work with integrals and derivatives. This is usually done in the derivation of mathematical equations where the dependence of quantities is predictable or even well known.

By the use of formulas (3.12) and (3.13), it can be proved that there exists a constitutive relation between the flux and the density in terms of $f = f(\rho)$.

Theorem 3.1 (Theorem 1 on p. 901 in [2]). *Let the arrival times $\tau(n,m)$ satisfy the time recursion* (3.4) *and let the approximate density ρ and flux f be defined by* (3.12), (3.13). *Then the approximate flux can be written in terms of the approximate density via a constitutive relation of the form*

$$f(\tau(n,m),x_m) = \min\left\{\mu(m-1), \frac{h_{m-1}\,\rho(\tau(n,m),x_{m-1})}{T(m-1)}\right\}, \quad n \geq 0, \quad m = 1,2,\dots.$$

Proof. The proof was done by Armbruster, Degond, and Ringhofer and can be found in [2]. $\square$

The second idea is now to investigate the continuum limit ($N, M \to \infty$) of (3.12) and (3.13). To be more precise, it is shown in which sense the approximate density and flux satisfy a conservation law of the form $\partial_t \rho + \partial_x f = 0$. The rather technical asymptotic validity can be divided into three parts: scaling, interpolation, and weak formulation. The starting point will be a dimensionless scaling where the reference value is given by the average processing time T_0 defined as

$$T_0 = \frac{1}{M}\sum_{m=0}^{M-1} T(m).$$

This choice of reference is easily explained since MT_0 describes the time a part takes to pass through the system without any waiting times. In the following, the average processing time is used as a scale basis over all time scales; i.e., all present variables and functions will be scaled by T_0 and labeled by the subindex s:

$$\tau(n,m) = MT_0\tau_s(n,m), \quad T(m) = T_0 T_s(x_m), \quad \mu(m) = \frac{\mu_s(\tau_s(n,m+1),x_m)}{T_0}. \tag{3.14}$$

Inserting (3.14) into the time recursion (3.4) we get for all n,m

$$\tau_s(n,m) = \max\left\{\tau_s(n+1,m) + \varepsilon T_s(x_m),\, \tau_s(n,m+1) + \frac{\varepsilon}{\mu_s(x_m)}\right\}, \tag{3.15}$$

where the initial and boundary conditions $\tau_s^A(n)$ and $\tau_s^I(m)$ are scaled in exactly the same manner. So the regime introduced is valid for a large number of suppliers $M \gg 1$ which could be also expressed by $\varepsilon = \frac{1}{M} \ll 1$. At this point, we neglect the scaled reformulation of the approximate flux (3.12) and density (3.13) since this can be done in a straightforward way. The main difficulty arising in the limiting process $M \to \infty$, or equivalently $\varepsilon \to 0$, is that the flux function can become discontinuous due to the assumption of different maximal capacities. Think of the following bottleneck situation (see Figure 3.5): If processor x_m has a higher capacity than processor x_{m+1}, i.e., $\mu(x_m) > \mu(x_{m+1})$, a queue in front of x_{m+1} will be built up. More precisely, delta distributions will occur in the density at this point since mass has to be conserved. Hence, the limiting density will be a distribution and not a classical function.

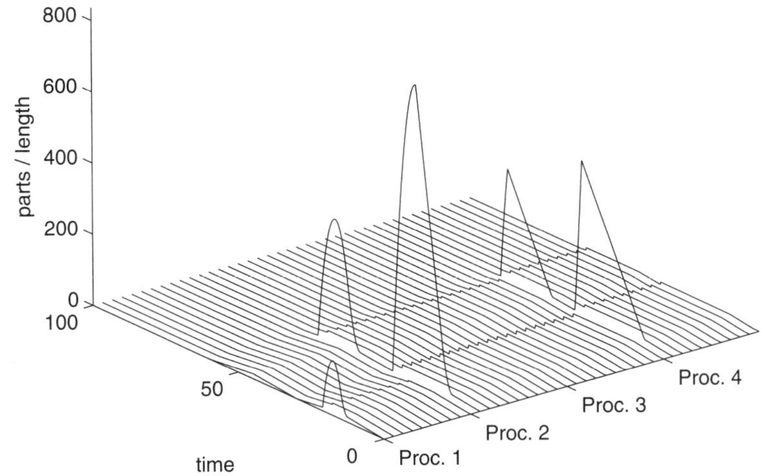

Figure 3.5. *Linear chain of four processors. Bottlenecks (exhibiting delta distributions) may appear only at those processors where the incoming flux is higher than its maximal processing rate.*

To prevent this issue, an asymptotic analysis for the N-curve $U(t,m)$ in (3.7) is performed. Let us therefore go into greater detail. The concept of upper bounds for the flux function, fixed by the capacities, allows for the assumption that the final conservation law will be of the form

$$\partial_t \rho + \partial_x f = 0, \quad f = \min\left\{\frac{L}{T}\rho, \mu(x)\right\}, \quad f(t,0) = f^A(t) \tag{3.16}$$

supplemented with initial conditions $\rho(0,x) = \rho_0(x)$. The relation between the density and the N-curve is now utilized to set $\rho(t,x) = -\partial_x u(t,x)$, where $u(t,x)$ is an approximation of $U(t,m)$. For simplicity, the processing velocity $V := L/T$ is chosen to be 1. Applying the relation above and integrating (3.16) once with respect to x we obtain the corresponding conservation law for the N-curve $u(t,x)$. However, this reformulation allows for shock solutions in u. Those will appear as delta distributions in the density ρ:

$$\begin{aligned} &\partial_t u(t,x) - \min\left\{\mu(x),\ -\partial_x u(t,x)\right\} = 0, \\ &\lim_{x \to 0-} u(t,x) = g^A(t), \quad \frac{d}{dt} g^A(t) = f^A(t). \end{aligned} \tag{3.17}$$

Although this equation is written in a continuous framework, we are aware of the fact that $u(t,x)$ is only a scaled grid function in space and time so far. An effective instrument towards a continuum is piecewise constant interpolation in space as well as in time. Note that the interpolants are still scaled and characterized by $\tilde{u}_s$ and $\tilde{f}_s$, respectively. The next theorem shows that the so-defined N-curve $\tilde{u}_s$ satisfies (3.17) weakly in x and t as $\varepsilon \to 0$.

Theorem 3.2 (Theorem 2 on p. 904 in [2]). *Given the scaled density and flux at the discrete points* $x_m, \tau_s(m,n)$, *as defined in* (3.15), *let the scaled throughput times* $T_s(x_m)$ *stay* *uniformly bounded; i.e., $h_m = O(\varepsilon)$ holds uniformly in m. Assume finitely many bottlenecks for a finite amount of time; i.e., let $\Delta_m \tau_s(n,m)$ be bounded for $\varepsilon \to 0$ except for a certain number of nodes m and a finite number of parts n, which stays bounded as $\varepsilon \to 0$. Then, for $\varepsilon \to 0$ and $\max h_m \to 0$ the interpolated N function and flux $\tilde{u}_s, \tilde{f}_s$ satisfy the initial boundary value problem*

$$\partial_t \tilde{u}_s = \tilde{f}_s, \quad \tilde{f}_s = \min\{\mu, -\partial_x \tilde{u}_s\}, \quad t > \tilde{\tau}_s^I(x), \quad 0 < x < X, \tag{3.18}$$
$$\tilde{u}_s(\tilde{\tau}_s^I(x), x) = 0, \quad \lim_{x \to 0-} \tilde{u}_s(t,x) = \int_{\tilde{\tau}_s(0,0)}^{t} f^A(s)\,ds$$

in the limit $\varepsilon \to 0$, weakly in x and t.

Proof. The proof was done by Armbruster, Degond, and Ringhofer and can be found in [2]. □

This theorem establishes the main result of this chapter. Starting from the time recursion (3.4) via the definitions for the approximate flux and density to the definition of interpolants, the theorem exhibits the asymptotic validity of the reformulated conservation law (3.18) for any N-curve u and flux function f. Considering unscaled variables, the theorem furthermore implies the approximate computation of the density via $\rho = -\partial_x u$.

Corollary 3.3. *In unscaled variables* (3.18) *can be rewritten as*

$$\partial_t u = \min\left\{\mu_-, -\frac{X}{MT_0}\partial_x u\right\}, \quad 0 < x < X, \tag{3.19}$$
$$\lim_{x \to 0-} u(t,x) = \int_{\tau(0,0)}^{t} f^A(s)\,ds, \quad \mu_- := \lim_{y \to x-} \mu(y),$$

where the solution is given by the unscaled N-curve $u(t,x)$.

The last thing missing in our discussion is the large number of suppliers required in the limiting process, i.e., $M \gg 1$. This assumption is not necessarily reasonable in many applications, but it is essential in considering the continuum limit. Anyhow, to reconcile both statements, one individual supplier is substituted by K so-called virtual processors. Theorem 3.4 exactly establishes the consistency of the exit times $\hat{\tau}(K,n)$ and $\tau(1,n)$ and consequently the generation of the same outflux. The proof is mainly based on the fact that bottlenecks can occur only in front of processors, corresponding to the first virtual processor. It remains to prove that all other virtual processors have empty queues, i.e., that the waiting times are zero $Q(n,m) = 0$; see Remark 3.2.1.

Theorem 3.4 (Theorem 3 on p. 907 in [2]). *Let the first processor S_0 in the chain be governed by* (3.4)*. If we replace the single processor by K virtual processors with the same processing rates and the same total throughput time, i.e., by*

$$\hat{\tau}(n+1,m+1) = \max\left\{\hat{\tau}(n+1,m) + \frac{T(0)}{K}, \hat{\tau}(n,m+1) + \frac{1}{\mu(m)}\right\}, \quad m = 0,1,\ldots,$$
$$\hat{\tau}(n,0) = \tau(n,0), \quad \hat{\tau}(0,m) = \tau(0,1) - \left(1 - \frac{m}{K}\right)T(0), \quad m = 1,2,\ldots,$$

then we obtain the same outflux; i.e.,

$$\hat{\tau}(n,K) = \tau(n,1), \quad n \geq 0,$$

holds.

Proof. The proof was done by Armbruster, Degond, and Ringhofer and can be found in [2]. □

A suitable number of virtual processors, necessary to ensure the asymptotic regime, is determined in possibly two ways: Either all processing times $\frac{T}{K}$ are roughly equidistributed or the total number of processors is enlarged by additional subprocessors.

Summarizing, as we have seen above, one can deduce a conservation law for the derivative $\rho(x,t) = -\partial_x u$. Because the resultant solution u allows for discontinuities, one obtains delta distributions on the level of the solution of the equation for ρ. These delta distributions (bottlenecks) are natural, modeling the queues in the system, but do not allow for a simple theoretical treatment of the equation.

3.4 Summary

We have introduced two models for production networks that are based on only a few assumptions and thus become applicable in many different settings. The formulation as a partial differential equation model is the method of choice in physics and engineering sciences to model time- and space-dependent relationships. Here, the main focus is on the derivation of a conservation law from a time recursion by mapping each supplier onto one gridpoint in space and performing an asymptotic analysis. This way of modeling is suitable in the case of simple networks without branches. The extension of the partial differential equation model to general network topologies essentially requires a new condition for the conservation of flow at branching points. This will be done by installing queues in front of each supplier; cf. Chapter 5. Then, as we will see, a major advantage is a new coupled system of equations which remedies the above bottlenecks in the density. The computational experiments in Chapter 8 as well emphasize the need of alternative modeling approaches in the context of supply chains and encourage thinking about other possible mathematical derivations from a microscopic (very detailed) to a macroscopic (rather coarse) way of description.

Chapter 4

Models Based on Ordinary Differential Equations

In this chapter we deal with ordinary differential equation models for supply chains and networks. More precisely, supplier production and good inventory evolutions are represented by ordinary differential equations, coupled through suitable demand-supply coefficients.

In the first section, we describe the system providing notation and equations for the network load dynamics. The specific case of supply chain is analyzed separately producing a dynamics much related to that of Chapter 3.

Then in Section 4.3, we describe the structure of the obtained network. Mathematically this amounts to a bipartite graph. The latter feature permits us to address optimal policy problems via the solution of a classical assignment problem.

Then we turn to the study of the global dynamics and relative steady states. The specific form of the system is exploited to get a characterization of steady states and also in the case of supply chains. Then we introduce the problem of linearized dynamics and bullwhip effect. The reader is referred to [44] for a detailed analysis. Finally, we consider the case of communicating suppliers and describe how the theory of consensus can be used profitably in this framework.

4.1 Motivation

Let us consider a real supply network as in Figure 4.1. There is a pine forest, from which two woodwork producers cut trees to obtain two different semifinished goods. Then the two woodwork producers sell their semifinished goods to two different factories. Finally the latter produce final goods, some of which are the same and some of which differ, placing them on the market.

Identifying the two producers with their semifinished goods, we can obtain a network with two goods (the two semifinished products) and two suppliers of final products. Moreover, we have flows from and to outside the network: the flow of pine trees from the forest to the two woodwork producers and the flow of final products from the factories to the market.

A discrete event model is possible, following all goods travelling on the network. However, if the quantities are big, then it is convenient to represent the status of the network assigning the level of flows and stored quantities of goods. Such information would be

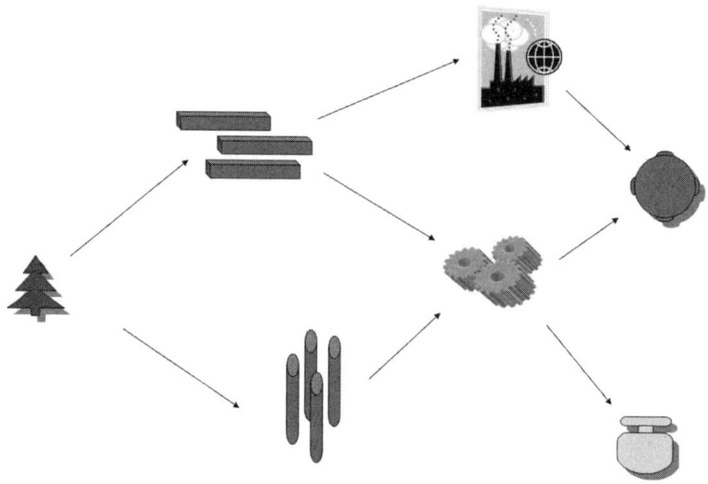

Figure 4.1. *An example of a real supply network.*

enough to manage production in an optimal way (regardless of the single good trajectories). Then the evolution of such quantities can be modelled by ordinary differential equations. More precisely, we will have variables indicating the level of inventories of goods and variables indicating the production levels. The interactions among such variables are obtained from the structure of the network with appropriate coefficients.

The final aim of such a model will be to understand possible oscillations in inventory levels and production rates and provide policies to tame them.

4.2 A Model for Supply Networks

The supply network model consists of J suppliers j producing I goods i. The suppliers receive goods from other suppliers, or from outside the network, and send the final products to other suppliers or directly to consumers. The production rate of the supplier j is indicated by μ^j and the inventory of the good i by q^i.

The whole dynamics is described by the inflows and outflows between the suppliers describing the network. More precisely, we have the following definition.

Definition 4.1. *We indicate by $j = 0$ a ghost supplier indicating the inflow into the network from the outside world and by $j = M + 1$ a ghost supplier indicating the outflow from the network to the outside world. The supply network dynamics is then given by the following:*

- *A stochastic matrix $D = (d_{i\,j})$ of deliveries, with $i = 1, \dots, I$, $j = 0, \dots, J$. More precisely, the outflow from supplier j to inventory q^i is given by $d_{i\,j}\mu^j$. We have $0 \le d_{i\,j} \le 1$ and $\sum_j d_{i\,j} = 1$.*

- *A stochastic matrix* $C = (c_{i\,j})$ *of consumptions, with* $i = 1,\dots,I,\ j = 1,\dots,J+1$. *More precisely, the inflow of good* i *needed by supplier* j *is given by* $c_{i\,j}\mu^j$. *We have* $0 \le c_{i\,j} \le 1$ *and* $\sum_j c_{i\,j} = 1$.

Notice that we indicate by $d_{i\,0}$ *and, respectively,* $c_{i\,J+1}$ *the inflow of resources from outside the network and the outflow of products to consumers outside the network. The term* $c_{i\,J+1}$ *may also take into account losses.*

We refer the reader to Figure 4.2 to illustrate Definition 4.1.

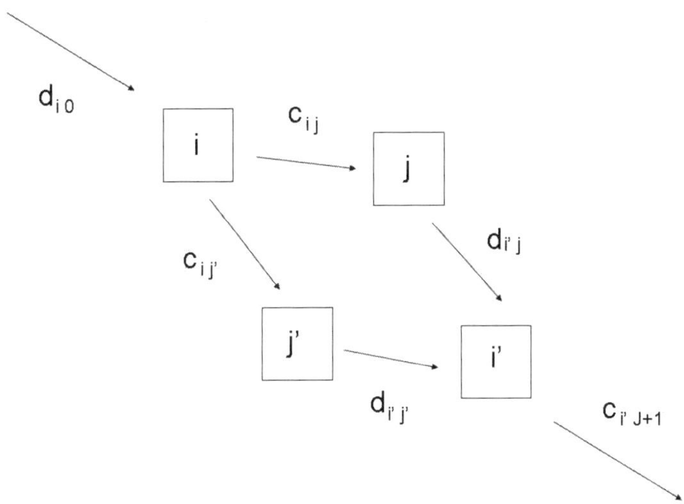

Figure 4.2. *A supply network.*

Interactions, by means of exchange of goods, are then represented by the supply matrix

$$S = D - C.$$

The inventory dynamics is then given by the following equation:

$$\frac{dq^i}{dt} = \left[\sum_{j=0}^{J} d_{i\,j}\,\mu^j\right] - \left[\sum_{j=1}^{J+1} c_{i\,j}\,\mu^j\right], \tag{4.1}$$

where the first term indicates the supply of the good (also from outside the network), while the second represents the demand. Thus the equation is nothing but a demand-supply balance.

On the other side, we expect the suppliers to adjust production rates depending on the level of inventories. More precisely, we assume that there exists a "desired production" function W_j, which assigns the appropriate production rate μ^j as a function of the inventories and possibly of their derivative. Then we write

$$\frac{d\mu^j}{dt} = -\frac{1}{T}\left[W_j\left(Q, \frac{dQ}{dt}\right) - \mu^j\right], \tag{4.2}$$

where T is a relaxation time, W_j is the desired production rate, and $Q = (q^1, \ldots, q^I)$ is the vector of inventory levels.

4.2.1 Supply Chain

A special case is that in which we have suppliers producing a single product; thus we identify them, and they are in a chain. In particular $I = J$, and we think that the inventory is placed before the supplier. Then $c_{j\,j} = d_{j\,j+1}$; thus we can represent the evolution giving only the matrix D for the sequence or supplier/inventory couples. More precisely, j receives goods from $j-1$ and sends them to $j+1$. Finally the evolution is given by (see Figure 4.3)

$$\frac{dq^j}{dt} = d_{j-1\,j}\mu^{j-1} - d_{j\,j+1}\mu^j, \tag{4.3}$$

$$\frac{d\mu^j}{dt} = -\frac{1}{T}\left[W_j\left(q^j, \frac{dq^j}{dt}\right) - \mu^j\right], \tag{4.4}$$

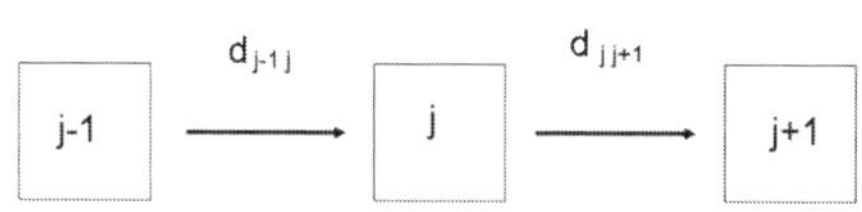

Figure 4.3. *Supply chain.*

4.3 The Structure of the Network

We are now interested in understanding the structure of the supply network. More precisely, we set the following definition.

Definition 4.2. *The graph associated with a supply network is given by the set of vertices* $\{1, \ldots, J+I\}$, *thus identifying suppliers with indices* 1 *to* J *and products with indices* $J+1$ *to* $J+I$. *Moreover, an edge* $(j, j+i)$ *exists if and only if* $c_{i\,j} \neq 0$ *or* $d_{i\,j} \neq 0$.

The next proposition immediately follows.

Proposition 4.3. *A supply network can be represented by a bipartite graph, with the set of suppliers and the set of inventories providing the partition.*

Indeed we have no edge between two suppliers or two inventories; see Figure 4.4.

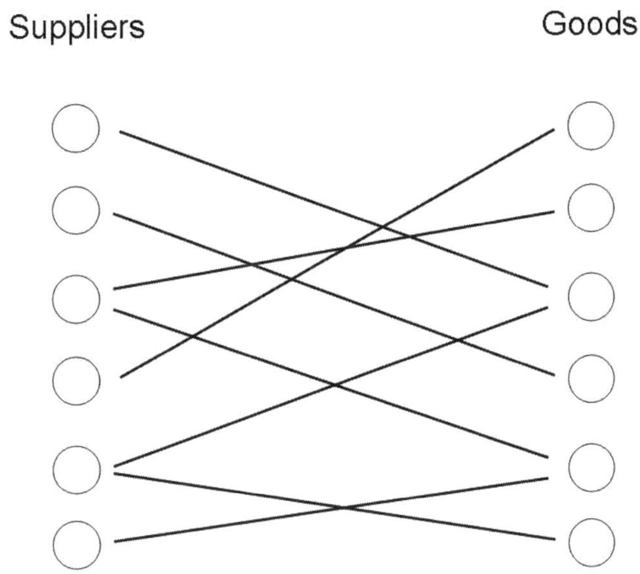

Figure 4.4. *The structure of the network: A bipartite graph.*

Bipartite graphs share many interesting properties. In particular, we may be interested in designing the network in an optimal way. In order to do that, consider the interaction terms

$$s_{i\,j} = d_{i\,j} - c_{i\,j},$$

which represents the exchange of goods between supplier j and inventory q^i. With the agreement $c_{i\,0} = 0 = d_{i\,J+1}$, we have

$$\sum_{j=0}^{J+1} s_{i\,j} = d_{i\,0} - c_{i\,J+1},$$

which represents the balance of the whole network regarding the good i.

We consider now the simplified situation in which only one of the two terms $c_{i\,j}$ and $d_{i\,j}$ can be different from zero. This amounts to saying that a supplier does not deliver to and consume from the same inventory at the same time. Also we assume that each supplier delivers to a single inventory and consumes from a single inventory (a reasonable situation in many practical cases). Then $c_{i\,j}, d_{i\,j} \in \{0,1\}$, and thus

$$|s_{i\,j}| \in \{0,1\}. \tag{4.5}$$

Normalizing the balances of goods and the productions we assume

$$\sum_j |s_{i\,j}| = \sum_i |s_{i\,j}| = 1. \tag{4.6}$$

If there is a specific return $R(i,j)$ by the exchange between the supplier j and the inventory q^i, then we can optimize the network structure solving a classical assignment problem, namely the following proposition.

Proposition 4.4. *Consider a supply network, with each supplier delivering to a single inventory, and vice versa, and assume also that a supplier cannot deliver to and consume from the same inventory.*

Then for a given matrix $R = (R(i,j))$, representing returns for exchanges between suppliers and inventories, after renormalization the optimization of the network is equivalent to the assignment problem:

$$\max \sum_{i,j} A(i,j)|s_{i,j}|$$

subject to (4.5) *and* (4.6).

We refer the reader to [14] for general results about the assignment problem.

4.4 Evolutions and Steady States

Let us indicate, for brevity, $M = (\mu^0, \ldots, \mu^{J+1})$ and, as above, $Q = (q^1, \ldots, q^I)$. Notice that D and C are $I \times (J+1)$ matrices. To express the evolution given by (4.1) and (4.2), we define two $I \times (J+2)$ matrices by

$$\widetilde{D} = \begin{pmatrix} 0 \cdots 0 \\ D \end{pmatrix}, \qquad \widetilde{C} = \begin{pmatrix} C \\ 0 \cdots 0 \end{pmatrix}, \qquad \widetilde{S} = \widetilde{D} - \widetilde{C}.$$

Then, in vector form we can write the evolution as

$$\dot{Q} = \widetilde{S}M, \tag{4.7}$$

$$\dot{M} = -\frac{1}{T}\left[W\left(Q, \widetilde{S}M\right) - M\right], \tag{4.8}$$

where $W = (W_0, \ldots, W_{J+1})$ with W_0 and W_{J+1} time varying and suitably chosen. Then the basic theory of ordinary differential equations provides existence and uniqueness of solutions as long as W is Lipschitz continuous with respect to Q and M and measurable in time.

4.4.1 Steady States

From (4.7) we immediately get that a steady state $(\bar{Q}, \bar{M})$ is characterized by

$$\widetilde{S}\bar{M} = 0, \qquad -\frac{1}{T}\left[W\left(\bar{Q}, \widetilde{S}\bar{M}\right) - \bar{M}\right] = 0,$$

from which we derive the following proposition.

Proposition 4.5. *A steady state $(\bar{Q}, \bar{M})$ satisfies*

$$\widetilde{S}W(\bar{Q}, 0) = 0, \qquad \bar{M} = W(\bar{Q}, 0).$$

In other words, $\bar{M} = W(\bar{Q}, 0)$ is an eigenvector of $\widetilde{S}$.

Steady States for Supply Chains

We now derive conditions for a steady state in the simplified case of a supply chain. Assume to have a steady state given by values $\bar{q}^j$ and $\bar{\mu}^j$. From (4.3), we derive

$$d_{j-1\,j}\bar{\mu}^{j-1} = d_{j\,j+1}\bar{\mu}^j,$$

which provides a link between the production rates of successive suppliers. Then from (4.3), we have

$$\bar{\mu}^j = W(\bar{q}^j, 0).$$

Again we can read the production rates as functions of the inventory levels, thus getting the following proposition.

Proposition 4.6. *Consider a supply chain and assume that the function $W(\cdot, 0)$ is invertible. Then a steady state is characterized by the relations*

$$\bar{q}^j = W(\cdot, 0)^{-1}\left(\frac{d_{j-1\,j}}{d_{j\,j+1}} W(\bar{q}^{j-1}, 0)\right),$$

$$\bar{\mu}_j = W(\bar{q}^j, 0).$$

In other words, for every $\bar{q}^0$ there exists a unique steady state.

4.4.2 Stability and Bullwhip Effect

The system (4.7) can be rewritten as a second-order system:

$$\ddot{Q} = -\frac{1}{T}\widetilde{S}\cdot\left[W\left(Q, \dot{Q}\right) - \dot{Q}\right]$$

in the variables Q, that is, inventory levels. Then one can use phase space analysis to study the stability of the system.

At an equilibrium $(\bar{Q}, \bar{M})$, we can linearize the policy W assuming

$$W(Q, S) = A\cdot(Q - \bar{Q}) + B\cdot(M - \bar{M}) + o(|Q - \bar{Q}|, |M - \bar{M}|).$$

Then the linearized dynamics in the variable $\tilde{Q} = Q - \bar{Q}$ reads

$$\ddot{\tilde{Q}} = -\frac{1}{T}\widetilde{S}\cdot\left[A\cdot\tilde{Q} - (B - I)\cdot\dot{\tilde{Q}}\right].$$

One can look at explicit solutions for this system. In particular, oscillations may be amplified, reproducing, for some choice of the parameters, the well-known bullwhip effect; see [44].

4.4.3 Consensus

An interesting case is that in which we have $J = I$ (for instance each supplier is producing a single good) and we look for consensus among different suppliers. Then we can consider a weighted communication graph and the relative symmetric stochastic weighted communication matrix $W = (w_{jk})$, with $w_{jk} \geq 0$ for $j \neq k$ and $w_{jj} = -\sum_{k \neq j} w_{jk}$. The meaning of W is the following: $w_{jk} \in [0,1]$ represents the communication level between supplier j and supplier k. Then we can use the protocol

$$W\left(Q, \widetilde{S}M\right) = M - W \cdot M. \tag{4.9}$$

What happens is that the obtained dynamics for each supplier is

$$\dot{M} = \frac{1}{T}\,[W \cdot M] = -\frac{1}{T}\,[L \cdot M],$$

where L is the Laplacian of the weighted communication graph. If such a graph is connected, then the system can be proved to be stable and to converge to a multiple of the eigenvalue $(1, \ldots, 1)$; see [68].

In other words, we have the following proposition.

Proposition 4.7. *Assume that* $I = J$ *and that the production policy is regulated by the matrix of a connected communication graph as in* (4.9). *Then the production rate vector* M *converges to a vector* $(\bar{\mu}, \ldots, \bar{\mu})$ *for some* $\bar{\mu} > 0$.

4.5 Summary

We have introduced a model for supply networks consisting of a finite set of variables representing good inventory levels and supplier production rates. Interactions among these variables are given by the network structure, while their evolution is ruled by ordinary differential equations.

Such an approach has the advantage of representing the network status in a quite compact way, still leaving the possibility of managing oscillations in production rates and inventory levels. This approach is particularly convenient in the case of a large number of parts travelling on the network but limited numbers of goods types and suppliers.

After introducing the main framework, we investigated the particular structure of the obtained graphs, which happen to be bipartite. Such a property links optimization of the network structure to a classical assignment problem.

Then steady states and stability are discussed both for the case of a supply network and that of a supply chain. Finally, we proposed using results from consensus theory to manage network flows.

Chapter 5

Models Based on Partial Differential Equations

The fundamental idea of this chapter is the extension of the supply chain model introduced in Chapter 3 to general network topologies. We make use of only a few general assumptions such that the framework is applicable to a variety of problems. In the following, we will see that the main new ingredient is a separate modeling of queues which can be achieved by specifying conditions at each vertex of the network to ensure conservation of mass through nodes. The latter is common in modeling problems with an underlying network structure and investigated for different applications, e.g., in the field of traffic flow [5, 22, 39, 58, 62], gas pipelines [6, 18, 19, 45], and telecommunications [25]. Considering the reformulated supply chain model this finally leads to a coupled system of partial and ordinary differential equations. In fact, we will see that the derived network model has several advantages and can be used for the purpose of simulation as well as optimization (Chapter 7). Within the newly proposed framework it is even possible to cover real-world networks having multiple incoming and outgoing arcs for each vertex. However, in the case of rather simple network geometries, one can show the consistency of the original and the network model; cf. Section 5.3. One further advantage of this recent point of view is the easy accessibility to the existence theory of the network problem; see Section 5.4. With a network model at hand, we permit the chance to include more reasonable features in the model—think of multicommodity or due-date production. In Section 5.5 such an extended network model is presented.

5.1 Motivation

We start introducing this section studying the manufacturing process of industrial goods where unfinished products are now transported between interlinked production steps; see Figure 5.1. Let us initially consider a single product flow and face the dynamics inside the individual production facilities. Our goal is to predict the time evolution of the whole system. The sample process can be stated as follows: At the beginning of the production, raw material is introduced into the system on supplier S_1, whereas the actual manufacturing takes place on suppliers S_2 to S_7.

Again, each processor has an upper limit of parts that can be handled simultaneously. To avoid congestion, each processor has its own buffering area (queue) where the parts

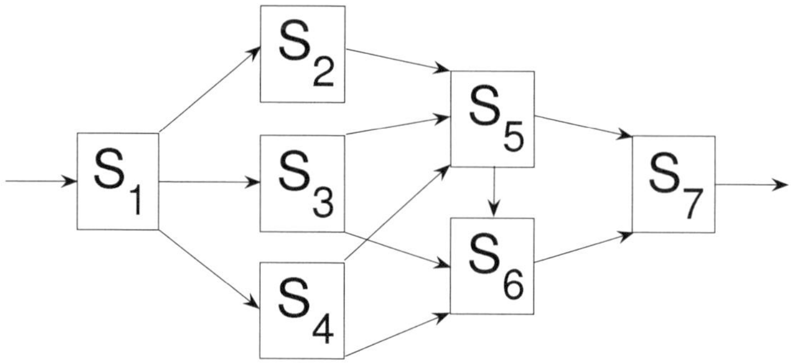

Figure 5.1. *Example of a production network.*

are possibly stored before the actual processing starts. Some questions which may arise in this context are the following: Can we simulate the evolution of the number of parts in the system? How do the previous values change if the buffering areas are reduced? Based on this kind of application we impose as few as possible assumptions on the underlying production process but still derive a physically valid and reasonable model. This type of model is a continuous model for the time and space evolution of a part density with particular attention to the dynamics inside each processor. However, we recover already known models by suitable approximations of our continuous model and furthermore present a mathematically constructive existence proof. The derived network model may serve as a basis for other models which might include further production aspects, e.g., scheduling of parts according to due dates.

5.2 General Network Topologies

This section is dedicated to the extension of continuous models to general network topologies. We therefore develop coupling conditions and discuss their necessity. In this context, we will also present a comparison of existing models.

To start working with network models it is essential to define a unique mathematical setting based on graph theory. In a first step, we give a definition for supply networks and introduce the basic notation used throughout the following chapters.

Definition 5.1 (Network definition). *A supply network is a finite, directed graph* $(\mathcal{A}, \mathcal{V})$ *where the set of arcs is given by* $\mathcal{A} = \{1, \ldots, M\}$. *Each arc* $e \in \mathcal{A}$ *of the network corresponds to an individual processor, and each vertex (node)* $v \in \mathcal{V}$ *represents the respective queue.*

1. *The length* L^e *of one processor is mapped on the interval* $[a^e, b^e]$. *We use* $a^{in} = -\infty$ *and* $b^{out} = +\infty$ *for the first (resp., the last) processor in the network.*

2. *According to the assumption that each processor possesses a queue, we locate a queue at the vertex* v *in front of the processor, i.e., at* $x = a^e$.

3. *The maximal capacity* μ^e *and the processing velocity* $V^e := L^e / T^e$ *of each processor are constant, user-defined parameters on each arc.*

Note that this definition is slightly different compared to common supply network graphs where processors are represented by nodes.

For a fixed vertex v, the set of ingoing arcs is denoted by δ_v^- and the set of outgoing arcs by δ_v^+. Let $x_v^e := a^e$ if $e \in \delta_v^+$, and let $x_v^e := b^e$ if $e \in \delta_v^-$. In the case of more than one outgoing arc, we introduce distribution rates $A^{v,e}(t)$; see, for instance, Figure 5.2. This corresponds to the same idea used in traffic flow (see [17, 33] for more details).

Definition 5.2. *For each vertex v with multiple outgoing arcs we introduce time-dependent distribution rates $A^{v,e}(t), v \in \mathcal{V}_d$, where $\mathcal{V}_d \subset \mathcal{V}$ denotes the set of dispersing intersections. The functions $A^{v,e}(t)$ are required to satisfy $0 \leq A^{v,e}(t) \leq 1$ and $\sum_{e \in \delta_v^+} A^{v,e}(t) = 1$ for all times $t > 0$.*

Those rates describe the distribution of the incoming flux among the outgoing processors and are later subject to optimization; see Chapter 7. Another possibility to model the distribution of parts is introduced in Section 5.4.2.

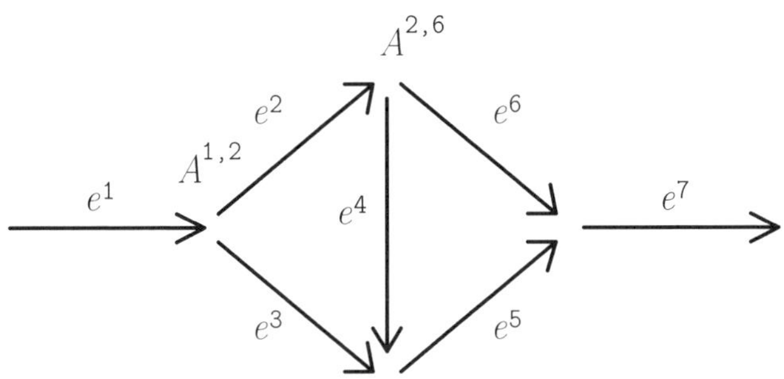

Figure 5.2. *Network with distribution rates $A^{1,2}(t)$ and $A^{2,6}(t)$.*

Before we pay attention to detailed application of distribution rates we focus on the different equations governing the flow on the network. In doing so, we briefly discuss the original transport equation introduced in Chapter 3 for one thing and derive the supply network equations step by step for another. More precisely, we recapitulate the equation

$$\partial_t \rho(t,x) + \partial_x f(\rho(t,x)) = 0, \tag{5.1a}$$

$$f(\rho(t,x)) = \min\left\{V(x)\rho(t,x),\ \mu(x)\right\}. \tag{5.1b}$$

Indeed, this equation can be interpreted as a supply network model describing the evolution of density along a straight line where processors are located at certain points in space. Note that the flux function is given by

$$f : \mathbb{R}_0^+ \times [0,X] \to [0,\mu(x)], \quad f(\rho,x) = \min\left\{V(x)\rho, \mu(x)\right\}, \tag{5.2}$$

with positive, piecewise constant functions $\mu(x), V(x)$ (see Figure 5.3). Clearly, f is Lipschitz continuous with Lipschitz constant $\theta = V(x)$.

We know that, whenever the flux is saturated, delta distributions in the density occur. This is always the case if incoming flux exceeds the maximal capacity μ^e at $x = a^e$. Therefore we introduce additional equations to our network model capturing the overload while we concentrate on defining suitable coupling conditions to connect all processors in the network and to guarantee the conservation of mass. We see that the conservation law (5.1) is reduced to a simple advection equation, while the new queue equations that buffer the surplus originally appeared as delta distributions. We now apply Definition 5.1 and see that the transport inside each processor e is governed by a simple advection equation with velocity $V^e > 0$ (see Figure 5.3). It holds for all $x \in [a^e, b^e]$, $t \in \mathbb{R}^+$ that

$$\partial_t \rho^e(t,x) + \partial_x f(\rho^e(t,x)) = 0, \tag{5.3a}$$

$$f(\rho^e(t,x)) = V^e \rho^e(t,x), \tag{5.3b}$$

where ρ^e denotes the density of parts at time t and point x on each arc e. In both models (5.1), (5.3), we require initial densities in the beginning for $t = 0$ and, furthermore, an inflow profile $f^{in}(t)$ as boundary condition for the first processor.

Remark 5.2.1. *Usually, an inflow profile $f^{in}(t)$ for the supply network is prescribed. In the above model, this profile can be transformed into initial data $\rho^{0,in}(x) := \rho^{0,in}(b^{in} - t) = f^{in}(t)$ on an (artificial) first arc where, corresponding to the assumption that $a^{in} = -\infty$, no queue exists. In addition we assume $\mu^{in} >$ max f^{in} and $V^{in} = 1$.*

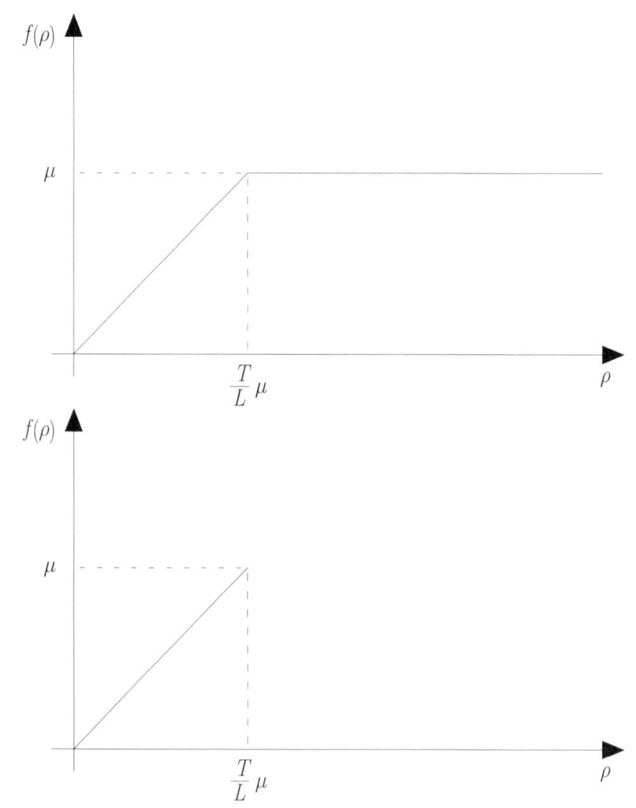

Figure 5.3. *The fundamental diagram of* (5.1) *and* (5.3), *respectively.*

Analogous to Definition 5.2, we define for a single vertex v the influx given by

$$A^{v,e}(t) \sum_{\bar{e}\in\delta_v^-} f^{\bar{e}}(\rho^{\bar{e}}(t,x_v^{\bar{e}})). \tag{5.4}$$

This is only one possibility to distribute parts. An alternative might be to consider rates $A^{v,e}(t)$ which additionally depend on external parameters, e.g., the priority of parts. As already mentioned, in contrast to (5.1), we consider queues in front of each processor which induces that the nonlinear conservation law (5.1) reduces to (5.3); cf. Figure 5.2. This fact can be realized by the following considerations.

Whenever a processor is connected to another processor of possibly different maximal capacity μ^e, we introduce a queue for the incoming but not yet processed parts. To describe the buffering we introduce the time dependent function $q^e(t)$ describing the load of the queue. Mathematically, we require each queue q^e to satisfy the following equation:

$$\partial_t q^e(t) = A^{v,e}(t) \sum_{\bar{e}\in\delta_v^-} f^{\bar{e}}(\rho^{\bar{e}}(t,x_v^{\bar{e}})) - f^e(\rho^e(t;x_v^e)), \tag{5.5}$$

where $f^e(\rho^e(t,x_v^e))$ denotes the outflow from queue e to processor e at time t and $q^e(0) = 0$ the initial condition. The dynamics of the buffering queue is governed by the difference of all incoming and outgoing fluxes at the connection point x_v^e. Corresponding to Definition 5.2, distribution rates are introduced to route the flux from preceding to successive processors. Let us consider different scenarios to point out the idea of defining distribution rates. Remember Figure 5.2 and assume arriving goods on arc e^1 shall continue on arc e^2 according to a given rate $A^{v,e}(t)$. Then, the rate $A^{1,2}(t)$ prescribes the percentage of parts going to e^2, and, consistent with Definition 5.2, the opposite rate $1 - A^{1,2}(t)$ determines the parts going to e^3. For simulation purposes it depends on the user whether to define the general time-dependent rates at dispersing junctions, i.e., $|\delta_v^+| > 1$, or to be satisfied with a stationary choice of them. Furthermore, considering the cases of solely one outgoing arc, i.e., $|\delta_v^+| = 1$, it should be remarked that the rates $A^{v,e}(t)$ become irrelevant since there is no freedom anymore to control the flux.

We describe the outflow of the queue into the processor by linking the incoming flux to the current status (empty or full) of the queue. If the queue is empty, the outgoing flux is either a percentage of the sum of all incoming fluxes given by $A^{v,e}(t)$ or the maximal processing capacity μ^e. In the first case the queue remains empty; in the second case the queue increases. Last, if the queue is full, the queue is always reduced with the maximal capacity:

$$f^e(\rho^e(t,x_v^e)) = \begin{cases} \min\left\{A^{v,e}(t)\left(\sum_{\bar{e}\in\delta_v^-} f^{\bar{e}}(\rho^{\bar{e}}(t,x_v^{\bar{e}}))\right), \mu^e\right\}, & q^e(t) = 0, \\ \mu^e, & q^e(t) > 0. \end{cases} \tag{5.6}$$

Numerically, there is a cutoff for some $q^e < \varepsilon$ since q^e is never exactly equal to zero. This modeling approach allows now for an exact computation of the original delta distributions in the density as illustrated in Figure 3.5. In order to avoid the appearing discontinuity in the queue length we refer the reader to [1], where a smoothed out version of (5.6) is introduced. One can show that

$$f^\varepsilon(\rho^\varepsilon(t,x_v^\varepsilon)) = \min\left\{\mu^\varepsilon, \frac{q^\varepsilon(t)}{\epsilon}\right\} \quad \text{with } \epsilon \ll 1 \tag{5.7}$$

is the approximate version of (5.6) which is a direct result from numerical calculus (cf. Section 5.3), relating the new coupled model to the averaged behavior of continuous problems. We will return to the reformulation when we concentrate on optimization problems; see Chapter 7.

Notice that (5.5) can be interpreted as Kirchhoff's law, where the sum of the outgoing flux plus the queue equals the incoming flux:

$$\sum_{e\in\delta_v^+} \left(\partial_t q^e(t) + f^e(\rho^e(x_v^e,t))\right) = \sum_{\bar{e}\in\delta_v^-} f^{\bar{e}}(\rho^{\bar{e}}(x_v^{\bar{e}},t)). \tag{5.8}$$

Hence, the network formulation is flexible and could be easily applied to diverse topologies. Due to the assumption that the dynamics on arcs is consistently described by ordinary and partial differential equations adding new arcs or nodes does not affect the nature of those equations; i.e., they still remain valid.

Remark 5.2.2. *The coupling conditions* (5.5), (5.6) *ensure furthermore that the network problem is well defined at inner nodes since they impose boundary conditions for the outgoing processor at* a^e.

We mention some examples occurring in a production process to highlight different network intersections as illustrated in Figure 5.4.

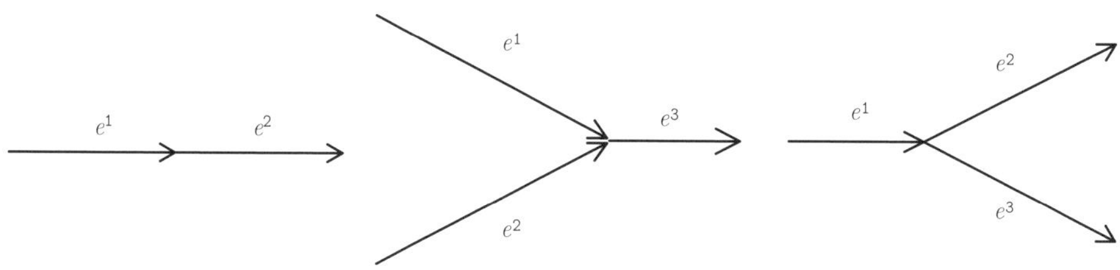

Figure 5.4. *Sample intersections labeled as types A, B, and C (from left to right).*

Sequence of Processors: Case A

First, we consider the case where only two arcs are connected. This is a vertex of degree two. Here, the incoming flux from e^1 is directly given into the queue of processor e^2. In the manufacturing sense think of some cans which are filled in processor e^1 and closed in processor e^2.

Merging Production System: Case B

Next, we discuss the merging of two production lines. We assume that the production in process e^3 does not need both e^1 and e^2 to be completed. Think, e.g., of a filling station for soda cans. Process e^3 fills the cans, whereas processes e^1 and e^2 produce plastic and aluminium cans, respectively.

Remark 5.2.3. *In contrast, the situation of some processes requiring both previous processors (think, e.g., of soda cans, where process* e^1 *delivers the can and* e^2 *delivers a label),* cannot *be modelled by a single queue only. Depending on the underlying production process the number of queues is given. In the case of merging cans and labels two different queues are needed. Then, the policy is to catch one can, then one label, and finally put them together. If one part is missing, both queues will grow up. Further, the number of cans and labels should be the same. This concept can be adapted to other situations but will not be considered throughout this book anymore.*

Distributed Production System: Case C

The opposite situation describes the splitting of the production line into two lines at a certain point. Here, distribution rates $A^{v,e}(t)$ are introduced for controlling the flux from the previous processor. Imagine a good produced on arc e^1 might retrieve a label in the English language on arc e^2 and in some other language on arc e^3. Assuming the arriving goods on arc e^1 continue on arc e^2 according to a given rate $A^{v,e}(t)$, e.g., finally 20% of all goods get a label in the English language. In convenience with Definition 5.2 the incoming flux on e^2 is given by $A^{1,2}(t)f^1(\rho^1(t,x_v^1))$ and on e^3 by $(1-A^{1,2}(t)(t))f^1(\rho^1(t,x_v^1))$. Summarizing, we obtain a system of partial differential equations governing the dynamics on each processor coupled to ordinary differential equations for the evolution of queues. The following equations hold true for $e \in \mathcal{A}$, $v \in \mathcal{V}$, $t \in (0,T)$ for all $x \in [a^e, b^e]$:

$$\partial_t \rho^e(t,x) + V^e \partial_x \rho^e(t,x) = 0, \tag{5.9a}$$

$$\partial_t q^e(t) = A^{v,e}(t) \sum_{\bar{e} \in \delta_v^-} f^{\bar{e}}(\rho^{\bar{e}}(t,x_v^{\bar{e}})) - f^e(\rho^e(t,x_v^e)), \tag{5.9b}$$

$$f^e(\rho^e(t,x_v^e)) = \begin{cases} \min\left\{A^{v,e}(t)\Big(\sum_{\bar{e}\in\delta_v^-} f^{\bar{e}}(\rho^{\bar{e}}(t,x_v^{\bar{e}}))\Big), \mu^e\right\}, & q^e(t) = 0, \\ \mu^e, & q^e(t) > 0, \end{cases} \tag{5.9c}$$

$$\rho^e(0,x) = \rho^{e,0}(x), \quad q^e(0) = q^{e,0}. \tag{5.9d}$$

To finally complete the modeling of the supply network model, in Section 5.3 we point out the correct mathematical relation between the model derived in [2] and the network formulation stated in (5.9).

5.3 Comparison to Existing Models

For a detailed analysis of the models presented in Sections 3.3 and 5.2, it will be sufficient to perform our numerical investigations on a network consisting of solely one processor. For simplicity, we therefore drop the superindex e. We show that discretizing and afterwards reformulating the nonlinear conservation law (5.1) yields the proposed network model (5.9) in the continuum limit.

At the first step, we start with the discretization of the processor using an equidistant grid with D points such that $\Delta x = x^i - x^{i-1}$, $i = 1, \ldots, D-1$, denotes the spatial grid size. The ongoing discussion is based on

$$\partial_t \rho + \partial_x f = 0, \quad f = \min\{V(x)\rho, \mu(x)\}, \quad f(0,t) = f^{in}(t), \tag{5.10}$$

with positive processing velocity $V(x)$. We will see, although the concept of solution of the network problem is different, that both models are equivalent. We also refer the reader to [50] for further comments.

Towards our comparison we apply a left-sided upwind discretization to (5.10). This is a reasonable discretization scheme since the flux function is monotone; i.e., the density is either carried with velocity V or stationary. The analysis we are interested in concentrates mainly on two different sets of discretization points: the left boundary x^0 and the remaining interior points. Those considerations combined with the limiting process will finally lead to a model that is similar to the supply network model but on a different time scale.

For brevity, we write $\rho_n^i = \rho(t_n, x_i)$, $q_n^1 = q^1(t_n)$, and $f_n^{in} = f^{in}(t_n)$. The processing velocity V and the capacity μ need not be discretized since they are constant in space and time along the processor. Applying the above upwind method yields in general

$$\rho_{n+1}^1 = \rho_n^1 - \frac{\Delta t}{\Delta x}(\min\{V\rho_n^1, \mu\} - f_n^{in}), \tag{5.11a}$$

$$\rho_{n+1}^i = \rho_n^i - \frac{\Delta t}{\Delta x}(\min\{V\rho_n^i, \mu\} - \min\{V\rho_n^{i-1}, \mu\}), \quad i \geq 2, \tag{5.11b}$$

where the left boundary condition is given by the influx f_n^{in} into the system.

Let us analyze the first discretization point x^0 at the left boundary. We discuss the situation where the influx f_n^{in} into the processor exceeds the maximal capacity μ; i.e., there is more mass flux arriving than the processor is able to process. Consequently, due to conservation of mass, delta distributions will occur in the density at this point. Further, the processor is forced to work at its maximum, and hence $f_n^1 = \mu$. Otherwise, if $f_n^{in} \leq \mu$ holds, the influx is simply propagated through the processor. Depending on the situations mentioned above, the flux f_n^1 is determined by the function $\min\{V\rho_n^1, \mu\}$. But this also induces that the flux f_n^i is always bounded by the maximal capacity μ for the remaining discretization points x^i, $i \geq 1$:

$$\rho_{n+1}^i = \rho_n^i - \frac{\Delta t}{\Delta x}(V\rho_n^i - \min\{V\rho_n^{i-1}, \mu\}) \quad \forall i \geq 1. \tag{5.12}$$

To answer the question how this problem is connected to the network model (5.9) we pick (5.11a), keep (5.12) untouched, and introduce the relation $q_n^1 = \Delta x \rho_n^1$. The latter can be regarded as the number of parts located in $[x^0, x^1)$. We furthermore make the substitution $\varepsilon := \Delta x / V$, where $\varepsilon \ll 1$, and end up with

$$\frac{q_{n+1}^1 - q_n^1}{\Delta t} = \left(f_n^{in} - \min\left\{\frac{q_n^1}{\varepsilon}, \mu\right\}\right). \tag{5.13}$$

This equation can be interpreted as follows: Assume a small queue q_n^1; then we obtain for $q_n^1 \leq \varepsilon\mu$ that the relaxed outgoing flux f_n^1 is given by $f_n^1 = f_n^{in} + \mathcal{O}(\varepsilon)$. In this case, the queue remains empty, except for a small error of $\mathcal{O}(\varepsilon)$. Conversely, for $q_n^1 > \varepsilon\mu$, the outgoing flux is determined by the maximum capacity of the processor, i.e., $f_n^1 = \mu$. A more accurate, but discontinuous, formulation for the computation of the queue outflow, i.e., $\varepsilon \to 0$, is given by (5.6). The difference between those approaches is numerically highlighted by Figure 5.5.

Example 5.3.1. *Consider the setting of an initially empty network, i.e.,* $q_0^1 = 0$, *where* $\varepsilon = \Delta t$ *and* $f_n^{in} < \mu$ *for all times n. Then, a small computation shows that* $q_{n+1}^1 = \Delta t f_n^{in}$ *holds, although the queue must be emptied. Obviously, the relaxed equation* (5.13) *always causes artificial queues that are of order* $\mathcal{O}(\varepsilon)$; *see also* [1].

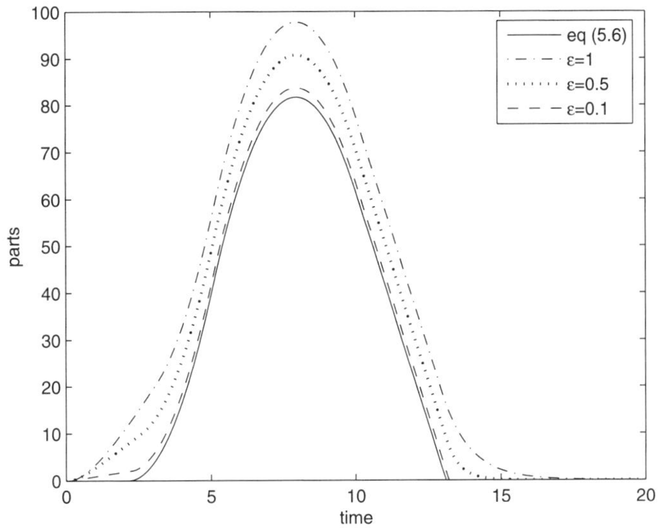

Figure 5.5. *A full queue computed via the original queue outflow* (5.6) *and its approximation* (5.13).

In the limit $\Delta t \to 0$, $\Delta x \to 0$ the above equations change to the continuous supply network model given by (5.3), (5.5), (5.6).

Altogether we can see that the definition of the network model derived above coincides with the conservation law in [2]. However, the definition of the density $\rho(t,x)$ is different. Here, in the network model, the points of discontinuity, i.e., the queues, are modeled by separate functions $q(t)$. Thus, the remaining part of $\rho(t,x)$ gives a well-defined density which can be treated by the usual front tracking algorithm avoiding the issue of distributional solutions.

5.4 Existence of Weak Solutions

In this section we discuss mathematical properties of the supply chain model introduced above. We prove local-in-time existence of a solution.

We construct approximate network solutions by the front tracking method; see [10, 21, 54]. Front tracking or wave tracking is a mathematical technique in order to establish existence results for hyperbolic partial differential equations. We briefly explain the basic ideas before giving the details. A proof of existence based on front tracking proceeds as follows. At first, we approximate the initial data by step functions, i.e., a series of initial Riemann problems, and the flux function by a piecewise linear function. If the flux function is piecewise linear, then all Riemann problems have solutions within the class of step functions. At some time $\bar{t} > 0$ one or more fronts may collide. We proceed by solving Riemann problems on each arc and at the vertices described in the previous sections. All wave interactions lead to new Riemann problems, either inside an arc or at a vertex. The new solution is again a step function with discontinuities travelling at constant speed. Therefore a global solution to the approximate problem can be obtained by tracking discontinuities and solving Riemann problems. This construction is well defined for all times t if the number of wave interactions is finite. It follows by a standard compactness argument that the approximate solutions converge to a network solution in L^1. The proof of Lipschitz dependence uses the approach as in [11].

We present the details of the proof. Since we deal with solutions to the advection and the queue equations we need a space coupling the dynamics of (ρ^e, q^e): density of parts and queue buffer occupancy. We prove existence and uniqueness of weak solutions for a general network of supply chains and bounded variation (BV) initial data. The densities ρ^e are Lipschitz continuous in time with respect to the L^1 metric, while the queue's buffer occupancies q^e are absolutely continuous. Moreover, we prove Lipschitz continuous dependence on the initial data. This, in turn, permits us to extend the corresponding semigroup trajectories to L^∞ initial data.

The dynamics for the part density $\rho^e(t,x)$ at supplier e, time t, and position x as introduced in Section 5.2 is given by

$$\partial_t \rho^e(t,x) + \partial_x f^e(\rho^e) = 0 \quad \forall x \in [a^e, b^e], \quad t \in \mathbb{R}^+, \tag{5.14a}$$

$$f^e(\rho^e) = \min\left\{V^e \rho^e, \mu^e\right\}, \tag{5.14b}$$

$$\rho^e(0,x) = \rho_0^e(x) \quad \forall x \in [a^e, b^e], \tag{5.14c}$$

$$\partial_t q^e(t) = A^{v,e} \sum_{\bar{e} \in \delta_v^-} f^{\bar{e}}(\rho^{\bar{e}}(t, x_v^{\bar{e}})) - f^e(\rho^e(t, x_v^e)), \tag{5.14d}$$

$$q^e(0) = q_0^e. \tag{5.14e}$$

If we compare (5.14) with the system (5.9) derived in the previous section, we observe the following: In the formulation (5.14) we kept the flux as $f = \min\{V\rho, \mu\}$. The reason is that for an initial value problem we cannot a priori assume that the density ρ_0^e on arc e is such that $V^e \rho_0^e$ is bounded by μ^e. In fact, we prove existence of weak solutions even in the case $V^e \rho_0^e > \mu^e$.

The analysis of the subsequent sections is conducted as follows. At first, we prove existence for a supply chain model which consists of a sequence of processors. We furthermore assume that the initial data is bounded by μ^e. Then, we extend the results to more general topologies, and finally we discuss the case of $\rho_0^e > \mu^e$.

We assume for the discussion following that

$$V^e(x) = V^e \textit{ is a constant.}$$

5.4.1 Part I—Consecutive Processors

First, we consider the case where each vertex is connected to exactly one incoming supplier and one outgoing supplier, and we assume that the arcs are consecutively labeled as in (5.16), i.e., arc e is connected to arc $e+1$, and that $b^e = a^{e+1}$; see also Figure 5.6.

In this case $A^{v,e} \equiv 1$ and equations (5.14) simplify to

$$\partial_t \rho^e(t,x) + \partial_x f^e(\rho^e) = 0 \quad \forall x \in [a^e, b^e], \quad t \in \mathbb{R}^+, \tag{5.15a}$$

$$f^e(\rho^e) = \min\left\{V^e \rho^e, \mu^e\right\}, \tag{5.15b}$$

$$\rho^e(0,x) = \rho_0^e(x) \quad \forall x \in [a^e, b^e], \tag{5.15c}$$

Figure 5.6. *Example of a simple network structure.*

$$\partial_t q^e(t) = f^{e-1}(\rho^{e-1}(t, b^{e-1})) - f^e(\rho^e(t, a^e)). \tag{5.16}$$

The outflux to the supplier is modelled as explained in the previous sections and given by

$$f^e(\rho^e(a^e, t)) = \begin{cases} \min\{f^{e-1}(\rho^{e-1}(t, b^{e-1}), \mu^e\}, & q^e(t) = 0, \\ \mu^e, & q^e(t) > 0. \end{cases} \tag{5.17}$$

To avoid technical difficulties, we assume that the first supplier ($e \equiv 1$) consists of a processor only and the last ($e \equiv M$) has infinite length, so $a^1 = -\infty$ and $b^M = +\infty$ for the first and, respectively, the last supplier in the supply chain.

Note that due to the shape of the flux function

$$f^e(\rho) := \min\{V^e \rho; \mu^e\} \tag{5.18}$$

a Riemann problem for (5.15) and $(t, x) \in \mathbb{R}^+ \times \mathbb{R}$ can be explicitly solved and admits rather simple solutions (see Figure 5.7):

Let $\rho_0^e(x) = \rho_l$ for $x < 0$ and $\rho_0^e = \rho_r$ for $x \geq 0$. If $\rho_l < \rho_r$, then the solution ρ^e is given by

$$\rho^e(t, x) = \begin{cases} \rho_l, & -\infty < \frac{x}{t} \leq \frac{f^e(\rho_r) - f^e(\rho_l)}{\rho_r - \rho_l}, \\ \rho_r, & \frac{f^e(\rho_r) - f^e(\rho_l)}{\rho_r - \rho_l} < \frac{x}{t} < \infty. \end{cases} \tag{5.19}$$

If, on the contrary, $\rho_l > \rho_r$, then the following happens. If either $\rho_l \leq \mu^e$ or $\rho_r \geq \mu^e$, then the solution is given by (5.19). Otherwise (i.e., if $\rho_r < \mu^e < \rho_l$), we obtain the solution given by

$$\rho(t, x) = \begin{cases} \rho_l, & -\infty < \frac{x}{t} \leq \frac{f^e(\rho_l) - \mu^e}{\rho_l - \mu^e}, \\ \mu^e, & \frac{f^e(\rho_l) - \mu^e}{\rho_l - \mu^e} < \frac{x}{t} \leq \frac{\mu^e - f^e(\rho_r)}{\mu^e - \rho_r}, \\ \rho_r, & \frac{\mu^e - f^e(\rho_r)}{\mu^e - \rho_r} < \frac{x}{t} < \infty. \end{cases} \tag{5.20}$$

Notice that the right-hand side of the first two inequalities is always 0 or 1.

A mathematical solution of the problem is given as follows.

Definition 5.3 (Network solution). *A family of functions* $\{\rho^e, q^e\}_{e \in \mathcal{A}}$ *is called an* admissible *solution for a network as in Figure* 5.6 *if, for all e,* ρ^e *is a weak entropic solution* [59] *to* (5.15), q^e *is absolutely continuous, and, in the sense of traces for* ρ^e*'s,* (5.16) *and* (5.17) *hold for almost every t.*

For the particular situation of a single vertex $v \in \mathcal{V}$ with incoming arc $e = 1$ and outgoing arc $e = 2$ and constant initial data $\rho_{e,0}(x) \leq \mu^e$, there exists an admissible solution

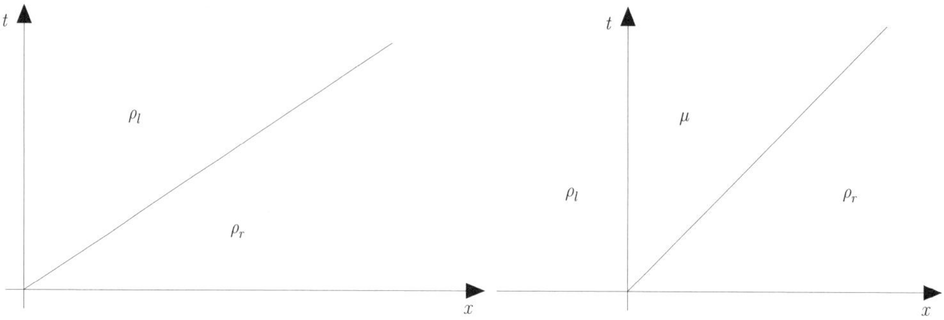

Figure 5.7. *The solution to the Riemann problem if $\rho_l < \rho_r$ (left) or if $\rho_r < \mu < \rho_l$ (right).*

$\{\rho^1, \rho^2, q^2\}$. The solution has the explicit form

$$\rho^1(t,x) = \rho_0^1, \tag{5.21a}$$

$$\rho^2(t,x) = \begin{cases} f^1(\rho_0^1) < \mu^2 & \begin{cases} \rho_0^1, & 0 \le (x-t_0)/t < 1 = \frac{f^2(\mu^2)-f^2(\rho_0^1)}{\mu^2-\rho_0^1}, \\ \mu^2, & 1 \le (x-t_0)/t \text{ and } x/t < 1, \\ \rho_0^2, & 1 \le x/t < \infty, \end{cases} \\ f^1(\rho_0^1) \ge \mu^2 & \begin{cases} \mu^2, & 0 \le x/t < 1 = \frac{f^2(\mu^2)-f^2(\rho_0^2)}{\mu^2-\rho_0^2}, \\ \rho_0^2, & 1 \le x/t < \infty, \end{cases} \end{cases} \tag{5.21b}$$

$$q^2(t) = q_0^2 + \int_0^t f^1(\rho_0^1) - f^2(\rho^2(a^2+,\tau))d\tau, \tag{5.21c}$$

wherein $t_0 = q_0^2/\left(\mu^2 - f^1(\rho_0^1)\right)$. In Figure 5.8, a sketch of the solution $\rho^2(x,t)$ in the x-t-plane in the case $q^{2,0} > 0$ and $f^1(\rho^{1,0}) < \mu^2$ is shown. For a network as in Figure 5.6, for initial data $\{\rho_0^e(x)\}^e$, where each ρ_0^e is a step function, and for initial values $q^e(0) = 0$, there exists an admissible solution $\{\rho^e, q^e\}^e$ to the network problem (5.15), (5.16), (5.17); see [38].

The construction of the solution is based on wave or front tracking (see below and [10, 20, 54]). For applications of this method in the context of network problems we also refer the reader to [33, 53].

To start, we introduce an equidistant grid $(i\delta)_{i=0}^{N_x}$ such that $0 \le (i\delta) \le \max\{\mu^e : e \in \mathcal{A}\}$ and such that for all e there exists i^e with $i^e = i\delta\mu^e$. Here, it is implicitly assumed that μ_i/μ^e is rational. We approximate the initial data by step functions $(\rho_0^\delta)^e$ taking values in the set $\{i\delta : i = 0, \dots, N_x\}$. Then each Riemann problem inside an arc or at a vertex is solved, obtaining various travelling discontinuities. If discontinuities collide, then the collision can be resolved either by solving a Riemann problem inside the arc e (see (5.19), (5.20)) or as a collision with a vertex (see (5.21)). In both cases we obtain new discontinuities propagating until the next collision.

At the same time, an evolution of the queue's buffers q^e is automatically defined when solving the Riemann problems at vertices.

This construction guarantees that the solution on arcs takes values only in the set $\{i\delta : i = 0, \dots, N_x\}$ and that we obtain a wave-front tracking approximate solution (denoted

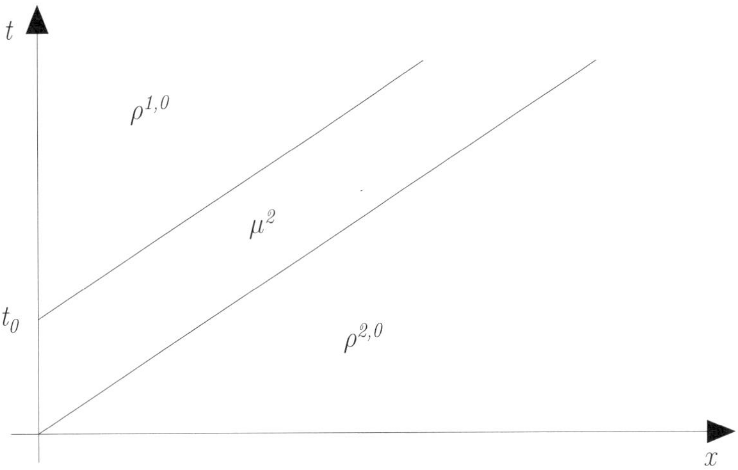

Figure 5.8. *Admissible solution at the vertex in the subcase* $q^{2,0} > 0$ *and* $f^1(\rho^{1,0}) < \mu^2$ *in the x-t-plane.*

by $(\rho^\delta, q^\delta) := \{(\rho_j^e, q_j^e)^\delta\}$) consisting of a set of moving discontinuities along the intervals $[a^e, b^e]$ and the queue's buffer evolutions.

As usual [10, 33], to guarantee the good definition of wave-front tracking approximate solutions and, passing to the limit, to prove existence of solutions in the sense of Definition 5.3, three basic estimates are in order:

1. the estimate on the number of waves,
2. the estimate on the number of interactions (between waves and of waves with queues), and
3. the estimate on total variation of solutions for ρ^e.

Moreover, in our case, we need to prove some compactness of the sequence $(q^e)^\delta$ in an appropriate space.

It is easy to check that every collision inside an arc decreases the number of waves, while the interactions with a vertex may produce two new waves; cf. (5.21). Also, since the characteristic velocity of waves is always positive and is bounded from above, the first two estimates are readily obtained; see [38]. Therefore the construction of wave-front tracking approximations is well defined up to any given time T.

Total Variation Estimates on Densities

Here, we provide total variation estimates on $(\rho^\delta)^e$ (i.e., along wave-front tracking approximate solutions). This will imply the existence of an admissible solution for BV initial data ρ_0^e.

First, we discuss the case of initial data ρ_0^e additionally satisfying the following assumption (K):

(K) For every e the initial datum satisfies $\rho_0^e \le \mu^e$.

The above construction guarantees that (K) remains valid for every time along wave-front tracking approximate solutions. Each $(\rho^\delta)^e(t,x)$ is a piecewise constant function in x and thus will define a number of constant states $(\rho^\delta)^e_j$, $e=1,\dots,M$, $j=1,\dots,N^e$, where we assume that $(\rho^\delta)^e(\cdot,a^e)=(\rho^\delta)^e_j$, and so forth. We define the total variation of the flux on the network as

$$T.V.(f(\rho^\delta))=\sum_{e\in\mathcal{A}}T.V.(f^e((\rho^\delta)^e(\cdot,t)))=\sum_{e\in\mathcal{A}}\sum_{j=1}^{N^e}|f^e((\rho^\delta)^e_j)-f^e((\rho^\delta)^e_{j+1})|. \quad (5.22)$$

Note that, thanks to assumption (K), a bound on $T.V.(f^e((\rho^\delta)^e(\cdot,t)))$ provides also a bound on $T.V.((\rho^\delta)^e(\cdot,t))$, since $(\rho^\delta)^e_j\le\mu^e$ for all j,e. Furthermore, $T.V.(f(\rho^\delta))$ does not increase when discontinuities collide *inside* an arc j; see [10]. Next, we discuss the collision of a discontinuity with a vertex.

Lemma 5.4. *Assume a single vertex with incoming arc $e=1$ and outgoing arc $e=2$. Furthermore, assume constant states ρ^e_0, $e=1,2$, at the vertex and consider a discontinuity colliding at time t_0. Denote the new solution at the vertex after the collision by $\bar\rho^e$. Assume no more collisions of discontinuities happen until t^*. Then, for all $t_0<t<t^*$,*

$$\sum_{e=1}^{2}T.V.(f^e(\rho^e(\cdot,t)))+|\partial_t q^2(t)|\le\sum_{e=1}^{2}T.V.(f^e(\rho^e(\cdot,t_0)))+|\partial_t q^2(t_0)|. \quad (5.23)$$

Proof. By construction the colliding discontinuity has to arrive from arc $e=1$, and therefore the total variation of the flux on this arc decreases by $|f^1(\bar\rho^1)-f^1(\rho^1_0)|$. On the outgoing arc $e=2$ we distinguish two cases. First, assume that $f^2(\rho^2_0)=f^1(\rho^1_0)$. Then, due to (5.16) we have $\partial_t q^2(t_0)=0$. If $f^1(\bar\rho^1)\le\mu^2$, then $f^2(\bar\rho^2)=f^1(\bar\rho^1)$ and (5.23) holds. If, on the other hand, $f^1(\bar\rho^1)>\mu^2$, then due to (5.17), $f^2(\bar\rho^2)=\mu^2$ and again (5.23) holds, since for $t>t_0$

$$\begin{aligned}|f^1(\bar\rho^1)-f^1(\rho^1_0)|&=|\mu^2-f^1(\rho^1_0)|+|f^1(\bar\rho^1)-\mu^2|\\&=|f^2(\bar\rho^2)-f^2(\rho^2_0)|+|\partial_t q^2(t)|.\end{aligned}$$

In the second case, we assume $f^2(\rho^2_0)=\mu^2$. Then,

$$|\partial_t q^2(t_0)|=|f^1(\rho^1_0)-\mu^2|,$$

and we distinguish two more subcases depending on whether the queue is increasing or decreasing after the collisions. First, assume $f^1(\bar\rho^1)\ge\mu^2$; i.e., the queue q^2 is increasing with

$$|\partial_t q^2(t)|=f^1(\bar\rho^1)-\mu^2$$

and

$$f^2(\bar\rho^2)=f^2(\rho^2_0)=\mu^2. \quad (5.24)$$

Inequality (5.23) still holds, since

$$|f^1(\rho_0^1) - f^1(\bar\rho^1)| + |\partial_t q^2(t_0)| \geq |\partial_t q^2(t)|$$

for $t > t_0$. Second, assume $f^1(\bar\rho^1) < \mu^2$; i.e., the queue q^2 is decreasing. Let $\bar t$ be such that $q^2(\bar t) = 0$. Then (5.23) holds since for $t < \min\{\bar t, t^*\}$

$$T.V.(f^2(\bar\rho^2(\cdot,t))) = 0$$

and

$$|f^1(\bar\rho^1) - f^1(\rho_0^1)| + |f^1(\rho_0^1) - \mu^2| \geq |\mu^2 - f^1(\bar\rho^1)| = |\partial_t q^2(t)|.$$

If $\bar t < t^*$, we obtain a new travelling discontinuity on the outgoing arc $e = 2$ for times $t > \bar t$ when the queue q^2 becomes empty: $f^2(\bar\rho^2(a^2+,t)) = f^1(\bar\rho^1)$ and $\partial_t q(t) = 0$ for $t > \bar t$. Then, (5.23) still holds, since

$$T.V.(f^2(\rho^2(\cdot,t))) + |\partial_t q^2(t)|$$

is constant for this interaction. This finishes the proof. $\square$

Summarizing, we conclude that for all $\delta > 0$ the following holds for all $t > 0$:

$$\sum_{e=1}^{M} T.V.((\rho^\delta)^e(\cdot,t)) + \sum_{e=2}^{M} |\partial_t (q^\delta)^e(t)| \tag{5.25a}$$

$$\leq \sum_{e=1}^{M} T.V.((\rho_0^e)^\delta(\cdot)) + \sum_{e=2}^{M} |\partial_t (q^\delta)^e(0)|,$$

$$(\rho^\delta)^e(t,x) \leq \max_e \mu^e \quad \forall e, x. \tag{5.25b}$$

Total Variation Estimates on Queue Buffers

Let us now estimate the total variation of $\partial_t q^e$.

Lemma 5.5. *Assume we have a single vertex with incoming arc $e = 1$ and outgoing arc $e = 2$ (of infinite length). Furthermore, assume we have constant states $\rho_0^e, e = 1,2,$ at the vertex and consider a discontinuity collision at time t_0. Denote the new solution at the vertex after the collision by $\bar\rho^e$. Assume no more collisions of discontinuities happen until t^*. Then, for all $t_0 < t < t^*$,*

$$T.V.(\partial_t q^2, [t_0,t]) \leq 2\,|f^1(\bar\rho^1) - f^1(\rho_0^1)| + |\partial_t q^2(t_0)|. \tag{5.26}$$

Proof. The interactions are clearly the same as those examined in Lemma 5.4.

First, assume that $f^2(\rho_0^2) = f^1(\rho_0^1)$. Then, due to (5.16) we have $\partial_t q^2(t_0) = 0$. If $f^1(\bar\rho^1) \leq \mu^2$, then $f^2(\bar\rho^2) = f^1(\bar\rho^1)$ and $\partial_t q^2(t) = 0$; thus (5.26) holds because the left-hand side vanishes.

If, on the other hand, $f^1(\bar{\rho}^1) > \mu^2$, then

$$|f^1(\bar{\rho}^1) - f^1(\rho_0^1)| = |f^2(\bar{\rho}^2) - f^2(\rho_0^2)| + |\partial_t q^2(t)| \geq |\partial_t q^2(t)|;$$

thus (5.26) holds because $\partial_t q^2(t_0) = 0$.

In the second case, we assume $f^2(\rho_0^2) = \mu^2$. Then,

$$\partial_t q^2(t_0) = f^1(\rho_0^1) - \mu^2, \quad \partial_t q^2(t_0+) = f^1(\bar{\rho}^1) - \mu^2. \tag{5.27}$$

If the queue is increasing after the interaction, then

$$T.V.(\partial_t q^2(t), [t_0, t]) = |f^1(\rho_0^1) - \mu^2 - (f^1(\bar{\rho}^1) - \mu^2)| = |f^1(\rho_0^1) - f^1(\bar{\rho}^1)|. \tag{5.28}$$

Second, assume $f^1(\bar{\rho}^1) < \mu^2$; i.e., the queue q^2 is decreasing. Let $\bar{t}$ be such that $q^2(\bar{t}) = 0$. For $t < \min\{\bar{t}, t^*\}$, (5.28) still holds; thus we conclude the case $t^* \leq \bar{t}$. If, on the contrary, $\bar{t} < t^*$, we obtain a new travelling discontinuity on the outgoing arc $j = 2$ for times $t > \bar{t}$ when the queue q^2 becomes empty: $f^2(\bar{\rho}^2(a^2+, t)) = f^1(\bar{\rho}^1)$ and $\partial_t q^2(t) = 0$ for $t > \bar{t}$. Then,

$$\begin{aligned} T.V.(\partial_t q^2(t), [t_0, t]) &= |\partial_t q^2(t_0) - \partial_t q^2(t_0+)| + |\partial_t q^2(t_0+) - \partial_t q^2(t)| \\ &\leq |f^1(\rho_0^1) - f^1(\bar{\rho}^1)| + |f^1(\bar{\rho}^1) - \mu^2| \\ &\leq 2\,|f^1(\rho_0^1) - f^1(\bar{\rho}^1)| + |\partial_t q^2(t_0)|. \quad \square \end{aligned}$$

We can now reason as follows. Define

$$\eta = \min_e |b^e - a^e|$$

as the minimum length of a supplier and set

$$(TV^e)^k = T.V.(f^e((\rho^e)^\delta(\cdot, k\eta))), \qquad (q^e)^k = \partial_t (q^e)^\delta (k\eta).$$

Then by Lemmas 5.4 and 5.5, we get

$$T.V.(\partial_t (q^e)^\delta, [k\eta, (k+1)\eta]) \leq 2(TV^{e-1})^k + (q^e)^k,$$

$$(q^e)^k + (TV^{e-1})^k + (TV^e)^k \leq (q^e)^{k-1} + (TV^{e-1})^{k-1} + (TV^e)^{k-1}.$$

Moreover, denoting by $(\widetilde{TV}^M)^k$ the variation in the flux produced on the last supplier by the queue $(q^M)^\delta$ on the time interval $[k\eta, (k+1)\eta]$, we get

$$(TV^1)^k \leq (TV^1)^0,$$

$$(q^M)^k + (TV^{M-1})^k + \widetilde{(TVM)}^k \leq (q^M)^{k-1} + (TV^{M-1})^{k-1} + (\widetilde{TV}^M)^{k-1}.$$

Therefore, summing up on e and k we get the following:

$$\sum_{e=2}^{M} T.V.(\partial_t (q^e)^\delta, [0, K\eta]) = \sum_{e=2}^{M} \sum_{k=0}^{K-1} T.V.(\partial_t (q^e)^\delta, [k\eta, (k+1)\eta])$$

$$\leq \sum_{e=2}^{M}\sum_{k=0}^{K-1}\left(2(TV^{e-1})^k+(q^e)^k\right) \leq K\sum_{e=2}^{M}\left(2(TV^{M-1})^0+(q^e)^0\right).$$

Restating, we have

$$\sum_{e=2}^{M} T.V.(\partial_t(q^e)^\delta,[0,K\eta]) \leq K\sum_{e=2}^{M}\left(2\,(T.V.(\rho_0^{e-1})^\delta(\cdot))+|\partial_t(q^\delta)^e(0)|\right) \quad \forall t. \qquad \square \tag{5.29}$$

Existence of a Network Solution for BV Initial Data

For existence of solutions, we consider the space of data (ρ,q) on the supply chain with the norm

$$\|(\rho,q)\| = \sum_e \|\rho^e\|_{L^1} + \sum_e |q^e|. \tag{5.30}$$

Then, we want to find a solution in the space $Lip([0,T],L^1((a^e,b^e)))$ for the ρ components and in the space $W^{1,1}([0,T])$ for the q components.

Due to the special flux function, we obtain discontinuities travelling with speed v at most equal to 1. Therefore, we have for $t_1 < t_2$ and every e

$$\int_{a^e}^{b^e} |(\rho^\delta)^e(t_1,x)-(\rho^\delta)^e(t_2,x)|dx \tag{5.31}$$
$$\leq T.V.((\rho^\delta)^e(t_1,\cdot))|t_1-t_2| + \int_{t_1}^{t_2} |f((\rho^e)^\delta(t,a^e))|\,dt.$$

The estimate (5.31) guarantees Lipschitz dependence with respect to time in L^1, while (5.25) ensures uniform BV bounds. Therefore, by using standard techniques [10, 70], one can show that for $\delta \to 0$ a subsequence of ρ^δ converges in L^1 provided that $T.V.(\rho_0^e(x))$ is bounded. Furthermore, the limit solution ρ^* is a weak entropic solution for (5.15).

Concerning q^e, we observe that $\partial_t q^e$ are of BV. Again by BV compactness, we have that $\partial_t q^e$ converges by subsequences in BV, in particular, almost everywhere and strongly in L^1. Thus q^e converges uniformly. Finally q^e converges by subsequences in $W^{1,1}$.

Remark 5.4.1. *Notice that we can pass to the limit using the uniform Lipschitz continuities of q^e. In fact, by definition, $Lip_t(q^e) \leq \max\{\mu^{e-1},\mu^e\}$. Thus we can pass to the limit, obtaining Lipschitz continuous functions with the same bound on the Lipschitz constant.*

Also, we can pass to the limit using estimate (5.25) *and the Ascoli–Arzelá theorem, but in that case we cannot guarantee that $\partial_t q^e$ is in BV and that q^e is in $W^{1,1}$.*

Consider now the case in which (K) is violated. For every e, the data entering the supplier from a^e satisfies (K). Consider the generalized characteristic $\pi^e(t)$ starting from a^e at time 0 and let τ^e (possibly $+\infty$) be the time in which it reaches b^e. We can divide the supplier into two regions:

$$A^e = \{(t,x) : x \leq \pi^e(t)\}, \qquad B^e = \{(t,x) : x > \pi^e(t)\};$$

see Figure 5.9.

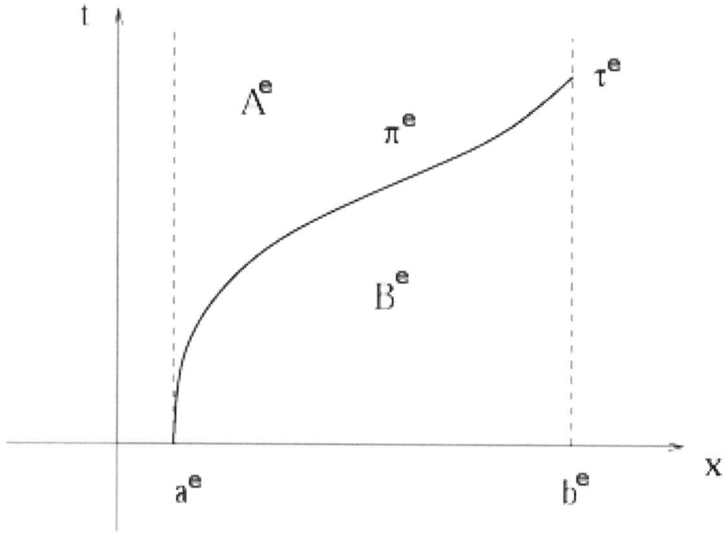

Figure 5.9. *Regions A^e and B^e.*

A^e is the region influenced by the incoming flux from a^e, while B^e is the region where ρ depends only on the initial datum ρ_0^e. Notice that for $t \geq \tau^e$, $B^e \cap \{(t,x) : a^e \leq x \leq b^e\} = \emptyset$. On A^e, (K) holds true; thus the estimate (5.25) also holds. On B^e the solution is the same as the solution to a scalar problem; thus the total variation is decreasing. We thus again reach compactness in BV and the existence of a solution.

Finally, we get the following proposition.

Proposition 5.6. *If $T.V.(\rho_0^e(x)) \leq C$ for some $C > 0$, then there exists a solution (ρ,q) on the network such that $(\rho,q) \in Lip([0,T], L^1((a^e,b^e))) \times W^{1,1}([0,T])$, ρ is BV for every time, and $\partial_t q^e$ is in BV.*

Uniqueness and Lipschitz Continuous Dependence

We want to prove uniqueness and Lipschitz continuous dependence on the space of data (ρ,q) on the supply chain with the norm (5.30). We use the same approach of [11, 33] and thus consider a Riemannian metric on this space where the tangent vectors are considered only for ρ^e piecewise constant functions.

Let us first focus on the ρ^e's: a "generalized tangent vector" consists of two components (v,ξ), where $v \in L^1$ describes the L^1 infinitesimal displacement, while $\xi \in \mathbb{R}^n$ describes the infinitesimal displacement of discontinuities. A family of piecewise constant functions $\theta \to \rho^\theta, \theta \in [0,1]$, with the same number of jumps, say, at the points $x_1^\theta < \cdots < x_M^\theta$, admits a tangent vector if the following functions are well defined (see Figure 5.10),

$$L^1 \ni v^\theta(x) \stackrel{\text{def}}{=} \lim_{h\to 0} \frac{\rho^{\theta+h}(x) - \rho^\theta(x)}{h},$$

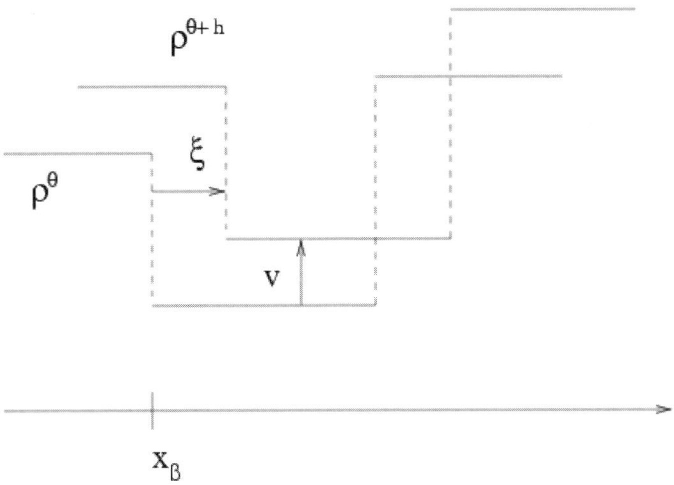

Figure 5.10. *Construction of "generalized tangent vectors."*

and also the numbers

$$\xi_\beta^\theta \stackrel{\text{def}}{=} \lim_{h\to 0} \frac{x_\beta^{\theta+h} - x_\beta^\theta}{h}, \qquad \beta = 1,\ldots,M.$$

Notice that the path $\theta \to \rho^\theta$ is not differentiable with respect to the usual differential structure of L^1; in fact, if $\xi_\beta^\theta \neq 0$, as $h \to 0$ the ratio $\left[\rho^{\theta+h}(x) - \rho^\theta(x)\right]/h$ does not converge to any limit in L^1.

The L^1 length of the path $\gamma : \theta \to \rho^\theta$ is given by

$$\|\gamma\|_{L^1} = \int_0^1 \|v^\theta\|_{L^1}\, d\theta + \sum_{\beta=1}^M \int_0^1 \left|\rho^\theta(x_\beta+) - \rho^\theta(x_\beta-)\right| \left|\xi_\beta^\theta\right| d\theta. \tag{5.32}$$

According to (5.32), the L^1 length of a path γ is the integral of the norm of its tangent vector, defined as follows:

$$\|(v,\xi)\| \stackrel{\text{def}}{=} \|v\|_{L^1} + \sum_{\beta=1}^M \left|\Delta\rho_\beta\right| \left|\xi_\beta\right|,$$

where $\Delta\rho_\beta = \rho(x_\beta+) - \rho(x_\beta-)$ is the jump across the discontinuity x_β.

Now, given two piecewise constant functions ρ and ρ', call $\Omega(u,u')$ the family of all "differentiable" paths $\gamma : [0,1] \to \gamma(t)$ with $\gamma(0) = u$, $\gamma(1) = u'$. The Riemannian distance between u and u' is given by

$$d(u,u') \stackrel{\text{def}}{=} \inf\left\{\|\gamma\|_{L^1}, \gamma \in \Omega(u,u')\right\}.$$

To define d on all L^1, for given $u,u' \in L^1$ we set

$$d(u,u') \stackrel{\text{def}}{=} \inf\left\{\|\gamma\|_{L^1} + \|u-\tilde{u}\|_{L^1} + \left\|u'-\tilde{u}'\right\|_{L^1} : \right.$$
$$\left. \tilde{u},\tilde{u}' \text{ piecewise constant functions, } \gamma \in \Omega(u,u')\right\}.$$

It is easy to check that this distance coincides with the distance of L^1.

To estimate the L^1 distance among wave-front tracking approximate solutions we proceed as follows. Take ρ,ρ' piecewise constant initial data and let $\gamma_0(\vartheta) = u^\vartheta$ be a regular path joining $\rho = \rho^0$ with $\rho' = \rho^1$. Define $\rho^\vartheta(t,x)$ to be a wave-front tracking approximate solution with initial data ρ^ϑ and let $\gamma_t(\vartheta) = \rho^\vartheta(t,\cdot)$. Then for every $t \geq 0$, γ_t is a differentiable path. If we can prove that

$$\|\gamma_t\|_{L^1} \leq \|\gamma_0\|_{L^1} \tag{5.33}$$

for every $t \geq 0$, then

$$\left\|\rho(t,\cdot) - \rho'(t,\cdot)\right\|_{L^1} \leq \inf_{\gamma_t} \|\gamma_t\|_{L^1} \leq \inf_{\gamma_0} \|\gamma_0\|_{L^1} = \left\|\rho(0,\cdot) - \rho'(0,\cdot)\right\|_{L^1}. \tag{5.34}$$

Now, to obtain (5.33), and hence (5.34), it is enough to prove that, for every tangent vector $(v,\xi)(t)$ to any regular path γ_t, one has

$$\|(v,\xi)(t)\| \leq \|(v,\xi)(0)\|; \tag{5.35}$$

i.e., the norm of a tangent vector does not increase in time. Moreover, if (5.34) is established, then uniqueness and Lipschitz continuous dependence of solutions to Cauchy problems are straightforwardly achieved passing to the limit on the wave-front tracking approximate solutions.

Remark 5.4.2. *Since the Riemannian distance d is equivalent to the L^1 metric, the reader could think that the whole framework is not so useful. On the contrary, the different differential structure permits one to rely on tangent vectors, whose norm can be easily controlled. This would not be possible using the tangent vectors of the usual differential structure of L^1, i.e., having only the v component.*

Also, while for systems of conservation laws it is possible to find a decreasing functional (see [12]*), this is not the case for networks (see* [33]*), even for a scalar conservation law.*

Let us now turn to the supply chain case. It is easy to see that all paths in L^1 connecting piecewise constant functions can be realized using only the ξ component of the tangent vector; see [11, 10]. Therefore, indicating by $x_{\beta_i^e}$ the positions of discontinuities, $e = 1,\ldots,M, i = 1,\ldots,M_e$, a tangent vector to a function defined on the network is given by

$$(\xi_{\beta_i^e}, \eta^e),$$

where $\xi_{\beta_i^e}$ is the shift of the discontinuity $x_{\beta_i^e}$, while η^e is the shift of the queue buffer occupancy q^e. The norm of a tangent vector is given by

$$\|(\xi_{\beta_i^e}, \eta^e)\| = \sum_{e,i} |\xi_{\beta_i^e}||\Delta\rho_{\beta_i^e}| + \sum_e |\eta^e|.$$

Again, to control the distance among solutions it is enough to control the evolution of norms of tangent vectors. Finally, we have the following lemma.

Lemma 5.7. *The norm of tangent vectors is decreasing along wave-front tracking approximations.*

Proof. The norm of tangent vectors changes only at interaction times or if a wave is generated (see [11]); thus we have to consider three cases.

(i) Two waves interact on a supplier.

(ii) A wave interacts with a vertex.

(iii) One queue empties down.

Case (i) is the same as the classical case; see [10, 11].

Consider case (ii) and assume that the interaction happens with vertex v at time t. Let us indicate by $f_e^{\pm}$ the value of the flux at a^e before and after the interaction and, similarly, by $(f^{e-1})^{\pm}$ the value of the flux at b^{e-1}. In general we use the symbols $+$ and $-$ to indicate quantities before and after the interaction, respectively. Assume first that $q^e(t) = 0$; then $(f^{e-1})^- = (f^e)^- < \mu^e$. If $(f^{e-1})^+ \leq \mu^e$, then the queue remains empty, a ρ wave is generated on supplier e, and the tangent vector norm remains unchanged. If $(f^{e-1})^+ > \mu^e$, then $\xi^+ = \xi^-$, $\Delta\rho^+ = \mu^e - (f^e)^-$, and $\eta^+ = \eta^- + \xi^-((f^{e-1})^+ - \mu^e)$. Since $\Delta\rho^- = (f^{e-1})^+ - (f^{e-1})^- = (f^{e-1})^+ - (f^e)^-$, the norm is conserved. Assume now that $q^e(t) > 0$; then $(f^e)^- = (f^e)^+ = \mu^e$. No ρ wave is produced, $\eta^+ = \eta^- + \xi^- \Delta\rho^-$, and again we conclude.

Let us pass to case (iii) and use the same notation of case (ii). Then $(f^e)^- = \mu^e$ and $(f^e)^+ = (f^{e-1})^- = (f^{e-1})^+ < \mu^e$. We get $\Delta\rho^+ = \mu^e - (f^{e-1})^-$, $\xi^+ = \eta^-/(\mu^e - f_{j-1}^-)$, and $\eta^+ = 0$; thus we are finished. $\square$

Existence for L^1 Initial Data

Since we proved Lipschitz continuous dependence, by an approximation argument, we also get existence for L^1 initial data. More precisely, we get the following theorem.

Theorem 5.8. *There exists a Lipschitz continuous semigroup S_t defined on the domain $\mathcal{D} = \{(\rho^e, q^e) : \rho^e \in L^\infty, q^e \in \mathbb{R}\}$. Moreover, for every initial datum (ρ^e, q^e) with ρ^e of BV, the semigroup trajectory $t \mapsto S_t(\rho^e, q^e)$ is a network solution.*

We point out that assumption (K) guarantees the existence of a solution on the network, while this is not granted in the general case, as shown by the following example.

Example 5.4.3. *Consider a simple network formed by only one vertex connecting an incoming arc $e = 1$ and an outgoing arc $e = 2$ and initial data*

$$\rho^e(0,x) = \mu^1 = \mu^2, \qquad \rho^2(0,x) = \mu^2 + \sin^2\left(\frac{1}{x - a^2}\right), \qquad q^2(0) > 0.$$

Clearly on the outgoing arc $e = 2$ the solution takes values in the flat part of the flux; thus it is constant in time. In particular, $\rho^2(t,x)$ has no trace as $x \to a^2$ for any value of t.

Remark 5.4.4. *Notice that* (5.17) *still makes sense for Example* 5.4.3 *if we interpret the relation to hold for every limit* $\lim_{x\to a^2} \rho^2(t,x)$. *On the other side, we can make oscillations in ρ^2 arbitrarily large if we put no constraints on the possible values of ρ^2.*

5.4.2 Part II—Junctions of Arbitrary Degree and Networks

Now, we turn to the case of more general networks as, for example, depicted in Figure 5.11.

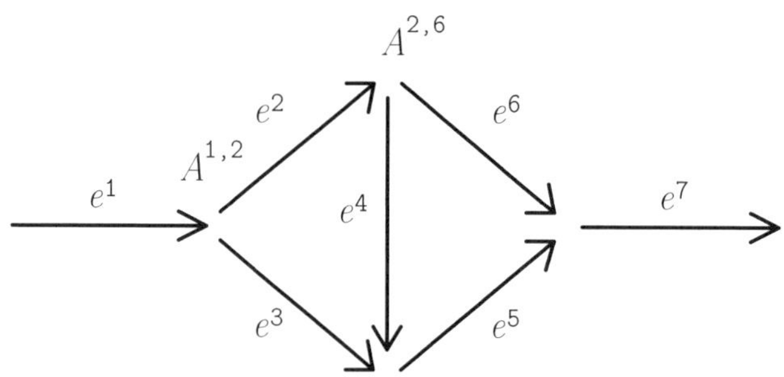

Figure 5.11. *Network geometry for a supply chain.*

We consider a more general case than in the previous section. We assume that for each single vertex v a matrix $(\alpha_{\tilde{e},e})_{\tilde{e},e} \in \mathbb{R}^{|\delta_v^-|\times|\delta_v^+|}$ is given and that the total flux willing to go to arc $e \in \delta_v^+$ is given by

$$\sum_{\tilde{e}\in\delta_v^-} \alpha_{\tilde{e},e} f^{\tilde{e}}(\rho^{\tilde{e}}(b^{\tilde{e}}-,t)).$$

The distribution rate introduced in the previous section is given by

$$A^{v,e} = \sum_{\tilde{e}\in\delta_v^-} \alpha_{\tilde{e},e}. \tag{5.36}$$

Therefore, we assume that the matrix $A^{v,e}$ satisfies, for all $e \in \delta_v^-$, $\tilde{e} \in \delta_v^+$, $0 \le \alpha_{\tilde{e},e} \le 1$ and $\sum_{e\in\delta_v^+} \alpha_{\tilde{e},e} = 1$. Then, the supply chain network model is given by (5.15) and for each junction v by the equations for the queues (see also [37]),

$$\forall e \in \delta_v^+ : \quad \partial_t q^e(t) = \sum_{\tilde{e}\in\delta_v^-} \alpha_{\tilde{e},e} f^{\tilde{e}}(\rho^{\tilde{e}}(b^{\tilde{e}}-,t)) - f^e(\rho^e(a^e+,t)), \tag{5.37}$$

and the boundary values for all $e \in \delta_v^+$,

$$f^e(\rho^e(a^e+,t)) = \begin{cases} \min\{\sum_{\tilde{e}\in\delta_v^-} \alpha_{\tilde{e},e} f^{\tilde{e}}(\rho^{\tilde{e}}(b^{\tilde{e}}-,t)); \mu^e\}, & q^e(t) = 0, \\ \mu^e, & q^e(t) > 0. \end{cases} \tag{5.38}$$

Note that due to the positive velocity of the occurring waves the boundary conditions are well defined. In particular, and in contrast to [33, 53], no additional maximization problem near the vertex has to be solved. Moreover, due to (5.54) and the assumption on A, we conserve the total flux at each vertex v for all times $t > 0$:

$$\sum_{e\in\delta_v^+} \left(\partial_t q^e(t) + f^e(\rho^e(a^e+,t))\right) = \sum_{\tilde{e}\in\delta_v^-} f^{\tilde{e}}(\rho^{\tilde{e}}(b^{\tilde{e}}-,t)).$$

Now, the construction of a solution to the network problem (5.15), (5.54), (5.38) is as before. In particular, the results of [38] extend to problem (5.15), (5.54), (5.38) on the network $(\mathcal{A},\mathcal{V})$. It is enough to control the number of waves and interactions: Let $\eta = \min_e(b^e - a^e)$ be the minimum length of a supplier. Since all waves move at positive velocity at most equal to 1, two interactions with vertices of the same wave can happen at most every η units of time. If N is the number of suppliers, then there is at most a multiplication by N every η units of time; thus we control the number of waves and interactions.

Therefore, for given piecewise constant initial data $(\rho^\delta)_0^e$ on a network, a solution (ρ^δ, q^δ) can be defined by the wave tracking method up to any time T. Next, we extend Lemma 5.4 to the more general situation of a vertex v above.

We can again get BV estimates.

Lemma 5.9. *Assume we have a single vertex with incoming arcs δ^- and outgoing arcs δ^+. Furthermore, assume we have constant states ρ_0^e, $e \in \delta^- \cup \delta^+$, at the vertex and consider a discontinuity collision at time t_0. Denote the new solution at the vertex after the collision by $\bar{\rho}^e$. Assume that there are no more collisions of discontinuities until t^*. Then, for all $t_0 < t < t^*$,*

$$\begin{aligned}&\sum_{e\in\delta^-\cup\delta^+} T.V.(f^e(\rho^e(\cdot,t))) + \sum_{e\in\delta^+} |\partial_t q^e(t)| \\ \leq &\sum_{e\in\delta^-\cup\delta^+} T.V.(f^e(\rho^e(\cdot,t_0))) + \sum_{e\in\delta^+} |\partial_t q^e(t_0)|.\end{aligned} \tag{5.39}$$

Proof. The proof is very similar to the proof of Lemma 5.4. The colliding discontinuity has to arrive on an arc $e \in \delta^-$, and we assume $e = 1$. The total variation on the incoming arc $e = 1$ therefore decreases by

$$|f^1(\bar{\rho}^1) - f^1(\rho_0^1)| = \sum_{e\in\delta^+} |\alpha_{1e} f^1(\bar{\rho}^1) - \alpha^{1e} f^1(\rho_0^1)|.$$

Hence, it suffices to prove that for any fixed outgoing arc $e \in \delta^+$ and for all $t > t_0$ the following inequality holds:

$$|\alpha_{1e} f^1(\bar{\rho}^1) - \alpha_{1e} f^1(\rho_0^1)| + |\partial_t q^e(t_0)| \geq T.V.(f^e(\rho^e(\cdot,t))) + |\partial_t q^e(t)|. \tag{5.40}$$

Fix $e \in \delta^+$. With the other cases being similar we discuss only the (most interesting) case: Assume

$$\sum_{\tilde{e}\in\delta^-} \alpha_{\tilde{e},e} f^{\tilde{e}}(\rho_0^{\tilde{e}}) > f^e(\rho_0^e)$$

and

$$\alpha_{1e} f^1(\bar{\rho}^1) + \sum_{\tilde{e},\tilde{e}\neq 1} f^{\tilde{e}}(\rho_0^{\tilde{e}}) < \mu^e.$$

Then, the queue q^e is decreasing after the collision at time t_0, and we denote again by $\bar{t}$ the time when $q^e(\bar{t}) = 0$. Then for $t < \min\{\bar{t}, t^*\}$ we obtain $T.V.(f^e(\bar{\rho}^e(\cdot,t))) = 0$ and $|\alpha_{1e} f^1(\bar{\rho}^1) - \alpha_{1e} f^1(\rho_0^1)| + |\partial_t q^e(t_0)| \geq \mu^e - \sum_{\tilde{e},\tilde{e}\neq 1} \alpha_{\tilde{e},e} f^{\tilde{e}}(\rho_0^{\tilde{e}}) - \alpha_{1e} f^1(\bar{\rho}^1) = |\partial_t q^e(t)|$. If $\bar{t} < t^*$, then a new discontinuity is generated since the queue q^e empties. By (5.38) we have

$$f^e(\bar{\rho}^e(a^e+,t)) = \sum_{\tilde{e},\tilde{e}\neq 1} \alpha_{\tilde{e},e} f^e(\rho_0^e) + \alpha_{1e} f^1(\bar{\rho}^1),$$

and therefore

$$|\partial_t q(\bar{t})| T.V.(f^e(\rho^e(\cdot,t)))$$

for $t > \bar{t}$. Hence, (5.40) holds for all $t > t_0$. This finishes the proof. □

Therefore, we again obtain the estimate (5.25), where the sum now should run over all arcs and nodes of the network. Moreover, the estimates on $\partial_t q^e$ work in the same way.

The same arguments as above give existence and uniqueness of a weak solution as well as the Lipschitz continuous dependence on the data in the general case for BV initial data. Finally, Theorem 5.8 holds for a general network.

5.5 Supply Chain Network Model

The basic supply chain model allows for extensions in different directions. We exemplify a procedure to extend the basic supply chain to more realistic situations. Currently, the supply chain model describes a production process where each part has the same properties. There is no distinction among these products. We extend this model to include additional attributes which products might have. Clearly, to obtain interesting models we need to assume that the attribute of a product has some influence on the dynamics of the supply chain, i.e., on the way it is processed. Hence, we consider in what follows a supply chain model where products have priorities. There exist some policies according to which products of different priorities are processed. Depending on these rules we obtain different dynamics. As an example for a priority rule one can think of the earliest due-date scheduling.

The starting point for deriving a model with policies will be a fine-scale kinetic equation for the probability distribution of finding a product of a certain priority at a certain place in space and time [3]. This kinetic framework allows one to model precisely the dynamics and will be the starting point for deriving supply chain models of macroscopic type as seen in the previous sections. In particular, the model which has been discussed throughout the previous sections will be obtained in the case of products with only one priority.

The derivation from kinetic to macroscopic supply chain models is closely related to the derivation of Euler's equation from kinetic Boltzmann models . We review the classical derivation of the system of gas dynamics from kinetic equations, which has been known in the literature [15] for a long time. We review these results since we apply a similar procedure below to a kinetic supply chain model. In classical rarefied gas dynamics the evolution equation for the density of particles $f(x,v,t)$ is given by

$$\partial_t f + v\partial_x f = Q(f,f) \tag{5.41}$$

for a velocity v and with a nonlinear collision operator $Q(f,f)$. This is formally given by

$$Q(f,f) = \int K(|v-w|)\big(f(x,v',t)f(x,w',t) - f(x,v,t)f(x,w,t)\big)\,dw\,dv.$$

Equation (5.41) is known as Boltzmann's equation. The collision operator enjoys the following properties:

$$\int Q(f,f)v\,dv = \int |v|^2 Q(f,f)\,dv = 0.$$

From these properties we deduce by integration from (5.41)

$$\rho_t + (\rho u)_x = 0$$

with $\rho = \int f dv$ and $(\rho u) = \int f v dv$. Similarly, by integration against v we obtain

$$(\rho u)_t + (\rho u^2)_x - \partial_x \sigma = 0$$

and by integration against $v^2/2$ we have

$$E_t + (Eu)_x - (\sigma u)_x + q_x = 0.$$

The other quantities are the higher-order moments defined as Cauchy stress tensor σ, internal energy e, total energy E, and heat flux q:

$$\sigma = -\int (v-u)^2 f\,dv, \quad \rho e = \int \frac{(v-u)^2}{2} f\,dv, \quad E = \frac{\rho u^2}{2} + \rho e, \quad q = \int \frac{(v-u)^3}{2} f\,dv.$$

From successive integrations we obtain a system of hyperbolic equations in the macroscopic variables $\rho, \rho u$, and E. Further integrations would yield additional equations. However, the equations are not closed since with any additional integration a new macroscopic variable has to be introduced. This cannot be solved without further assumptions. This is known as the closure problem. In classical gas dynamics the closure is typically taken as solution f^e to the stationary Boltzmann equation independent of time and space, i.e., $Q(f^e, f^e) = 0$. The function f^e is called Maxwellian, and using f^e the Euler and Navier–Stokes equations can be derived.

In the subsequent derivation this approach is mimicked in the following way. Starting from a kinetic equation we derive by integration the macroscopic moment models. The arising system of hyperbolic equations is not closed, and suitable closure relations have to be applied.

5.5.1 Kinetic Model for a Single Processor

The starting point for our discussion is the kinetic supply chain model for a single line including policy attributes [3]. Also, in this setting the supply chain consists of a sequence of queues and processors with certain throughput times and capacities. But now, the dynamics for new internal variables (priorities/attribute of the product) is given by a kinetic-like model. It is desirable to define policies according to the importance (given, e.g., by the due date) of the arriving parts due to the limited processor capacities. A simple policy would be as follows:

Parts labeled with an upcoming due date have to be served first, while the remaining parts have to wait.

The kinetic density is denoted as $f(x,y,t)$ at time t, position x, and attribute y, and further its evolution is generally governed by a Boltzmann-type equation. A free movement of parts is then given by

$$\partial_t f(x,y,t) + \partial_x[V(x,y,t)f(x,y,t)] = 0, \quad x \in \mathbb{R}, \quad y \in \mathbb{R}, \quad t > 0, \tag{5.42}$$

where $V(x,y,t)$ denotes the velocity function determined by the production process. Since the total flux $q(t,x) = \int_{\mathbb{R}} V(x,y,t)f(x,y,t)dy$ is bounded by the maximal capacity $\mu(x)$, i.e., $q(t,x) \leq \mu(x)$, and the attributes are ordered inversely $y_1 < y_2$, i.e., a part labeled by the attribute y_1 is processed before a part with attribute y_2, a natural choice for the velocity is

$$V(x,y,t) = v(x,y,t)H(\mu(x) - \beta(x,y,t)), \tag{5.43}$$

where $H(\cdot)$ denotes the Heaviside function. The computation of the threshold attribute $\beta(x,y,t)$ [3] is as follows: We take the product flux for all products having priority y' being higher than y, i.e.,

$$\beta(x,y,t) = \int H(y-y')v(x,y',t)f(x,y',t)\,dy'.$$

Obviously, due to (5.43), parts with an attribute less than $\beta(x,y,t)$ (high priority parts) move with velocity $v(x,y)$, while parts with attributes larger than $\beta(x,y,t)$ (low priority parts) are stopped, i.e., $V(x,y,t) = 0$. Thus, the general kinetic equation (5.42) changes to

$$\partial_t f + \partial_x[H(\mu(x) - \beta(x,y,t))v(x,y,t)f] = 0, \tag{5.44a}$$

$$\beta(x,y,t) = \int H(y-y')v(x,y',t)f(x,y',t)\,dy'. \tag{5.44b}$$

Note that the attributes are ordered inversely $y_1 < y_2$; i.e., a part labeled by the attribute y_1 is processed before a part with attribute y_2. More generally, lower values of y correspond to higher priorities. The kinetic model (5.44) provides a detailed description of the high-dimensional phase space. To reduce complexity of the model we consider higher-order moments of (5.44). As seen in [3], moment equations are obtained by integrating (5.44a) against powers of the attribute variable y^j for $j = 0,\dots,J-1$:

$$\partial_t m_j + \partial_x F_j = 0, \tag{5.45}$$

where the moments m_j and the moment fluxes F_j are given by

$$m_j(t,x) = \int y^j f(x,y,t)dy, \tag{5.46a}$$

$$F_j(t,x) = \int y^j H(\mu(x) - \beta(x,y,t))v(x,y,t)f(x,y,t)\,dy. \tag{5.46b}$$

Note that the jth moment system is an approximation for (5.44) stopped at a fixed j. Obviously, there are J equations (5.45) for $2J$ unknown variables m_j and F_j, $j = 0,\ldots,$ $J-1$. Since the moment system is not closed, a closure relation must be imposed to find additional J equations. We offer the following motivation to choose an ansatz as presented below: The necessity of using a smoothed-out version of δ is due to the fact that we need to integrate the product of δ and Heaviside functions. The real variable y defines a processing priority, and $\rho(t,x) = f(x,y,t)dy$ are the number of parts with priorities $[y, y+dy]$. We assume that the attribute distribution of a supply chain consists of K properties Y_k, $k = 1,\ldots,K$, with densities ρ_k. A particularly simple situation called single-phase flow is the case $K = 1$; i.e., all parts have the same property Y_1. Therefore, the following equilibrium is proposed to close the system (5.45), (5.46):

$$f^{\epsilon}(x,y,t) = \sum_{k=1}^{K} \rho_k(t,x)\frac{1}{\epsilon}\phi'\left(\frac{y - Y_k(t,x)}{\epsilon}\right), \tag{5.47}$$

where $\frac{1}{\epsilon}\phi'(\frac{y-Y_k(t,x)}{\epsilon}) \to \delta(y - Y_k(t,x))$, $\epsilon \to 0$ or, to be more precise, $\phi(y)$ is smooth such that $\phi'(y)$ is compactly supported and $\int \phi'(y)dy = 1$.

Next, we recall macroscopic equations obtained from (5.45)–(5.47) for the multiphase model $K \geq 1$. The moment equations are obtained by multiplying the kinetic equation by y^j and integration on $\int_{-\infty}^{\infty}$ with respect to y. Introducing the moments $m_j = \int y^j f(x,y,t)dy$ we obtain the following proposition. A proof can be found in [3].

Proposition 5.10 (Propositions 3.1 and 3.2 in [3]). *Using the ansatz* (5.47) *the moment fluxes* (5.46) *are given asymptotically by*

$$m_j(t,x) = \sum_{k=1}^{K} \rho_k(t,x)Y_k^j(t,x) + O(\epsilon), \tag{5.48a}$$

$$F_j(t,x) = \sum_{k=1}^{K} \rho_k(t,x)Y_k^j(t,x)v(x,Y_k,t)Z_k(t,x) + O(\epsilon), \tag{5.48b}$$

$$Z_k(t,x) = \max\left\{0, \min\left\{1, \frac{\mu(x) - \sum_{Y_s \neq Y_k} \rho_s(t,x)v(x,Y_s,t)H(Y_k - Y_s)}{v(x,Y_k,t)\sum_{Y_s = Y_k} \rho_s(t,x)}\right\}\right\}. \tag{5.48c}$$

Consider (5.45), *let* $J = 2\,K$, *and let*

$$m_j(t,x) = \sum_{k=1}^{K} \rho_k(t,x)Y_k^j(t,x),$$

$$F_j(t,x) = \sum_{k=1}^{K} \rho_k(t,x)v(x,Y_k,t)Z_k(t,x)Y_k^j(t,x).$$

Then (5.45), *for* $j = 0, \dots, J-1$, *is equivalent (at least for smooth solutions) to*

$$\partial_t \rho_k + \partial_x q_k = 0, \quad k = 1, \dots, K, \tag{5.49a}$$

$$\partial_t (\rho_k Y_k) + \partial_x (q_k Y_k) = 0, \quad k = 1, \dots, K, \tag{5.49b}$$

$$q_k(t,x) = \rho_k(t,x) v(x, Y_k, t) Z_k(t,x) \tag{5.49c}$$

as long as Y_k *are mutually distinct.*

A more explicit form of (5.49) in the case of single- and two-phase flows is derived in [3] and reformulated in the following remark for convenience.

Remark 5.5.1. *We define* $v_k := v(x, Y_k, t)$ *and* $q_k := \rho_k v_k Z_k$. *In the case of single-phase flow* ($K = 1$), (5.49) *simplifies to*

$$\partial_t \rho + \partial_x (\min\{\mu, \rho v_1\}) = 0, \tag{5.50a}$$

$$\partial_t (\rho Y_1) + \partial_x (Y_1 \min\{\mu, \rho v_1\}) = 0. \tag{5.50b}$$

If $v_1 = v(x)$, *then the above equations decouple and we recover the model proposed in* [2, 4].

The discussion is slightly more involved in the case of two-phase flow ($K = 2$). *We assume* $Y_1 < Y_2$, *i.e., parts with lower values of* Y_k *have larger priorities, and obtain for* $k = 1, 2$

$$\partial_t \rho_k + \partial_x q_k = 0, \tag{5.51a}$$

$$\partial_t (\rho_k Y_k) + \partial_x q_k Y_k = 0, \tag{5.51b}$$

with the following definition of q_k:

1. *If* $\mu < \rho_1 v_1$, *then* $q_1 = \mu$ *and* $q_2 = 0$.
2. *If* $\rho_1 v_1 < \mu < \rho_1 v_1 + \rho_2 v_2$, *then* $q_1 = \rho_1 v_1$ *and* $q_2 = \mu - \rho_1 v_1$.
3. *If* $\rho_1 v_1 + \rho_2 v_2 \leq \mu$, *then* $q_k = \rho_k v_k$ *for* $k = 1, 2$.

We offer the following interpretation of these cases: If the flow of parts of priority Y_1 *is already exceeding the maximal capacity* μ, *then we process only these parts and do not move the others (case* 1*). If we still have some capacity left after processing all parts of higher priority, we process as much as possible of the lower capacity part* $q_2 = \mu - \rho_1 v_1$ *(case* 2*). Finally, if there is sufficient capacity to process parts of either priority, we do so (case* 3*). Note, if* $Y_1 > Y_2$, *that the roles of* q_1 *and* q_2 *will be exchanged.*

As already observed, even the single-phase flow independent of attributes allows for delta solutions since the flux function can become discontinuous due to rapid changes in the capacity $\mu(x)$. This drawback is removed by proposing a network formulation of single-phase flow with attribute-independent velocity. In the following we want to use the network approach and to combine this with the multiphase model introduced above. This can be accomplished by first extending the kinetic model to a network and then, in a second step, deriving corresponding moment equations in the network formulation. This allows for a treatment of an arbitrary number of phases in the final network model. Note that in the network model for multiple policies we still have delta solutions due to attribute-dependent velocities. To simplify the discussion we first consider only a simple chain as the prototype for a network and comment on the changes for complex interconnected networks. Examples of the final network equations will be given for the single-phase and two-phase flow.

The basic idea of a network formulation of a simple chain of suppliers is as follows: We assume that the processing capacity μ is piecewise constant for each processor such that

$$\mu(x) = \sum_{e \in \mathcal{A}} \mu^e \chi_{[a^e, b^e]}(x). \tag{5.52}$$

Here χ is the characteristic function. Under this assumption, it is reasonable to split the full model (5.44) according to the jump discontinuities in μ. This induces a chain structure where each processor is related to a constant, fixed processing rate μ^e, according to our network definition (Definition 5.1). Different processors are then coupled at vertices (of degree at most two) by suitable coupling conditions; see below. Moreover, we might extend the previous discussion to allow for vertices of degree larger than two, which corresponds to interconnected processors.

5.5.2 Kinetic Model on a Network

On each arc $e \in \mathcal{A}$, the dynamics is governed by the kinetic equation (5.44) for f^e, the density of parts on arc e, and fixed, constant processing capacity μ^e:

$$\partial_t f^e + \partial_x [H(\mu^e - \beta^e(x,y,t)) v^e(x,y,t) f^e] = 0, \quad x \in [a^e, b^e], \quad t > 0, \tag{5.53a}$$

$$\beta^e(x,y,t) = \int H(y - y') v^e(x,y',t) f^e(x,y',t)\, dy'. \tag{5.53b}$$

Next, we introduce coupling conditions for connected processors. Since the processors may have different capacities μ^e, we need to introduce a buffer for parts which cannot be immediately processed. Consider a chain-like network (see Figure 5.12), i.e., $\delta_v^+ = \{e\}$ and $\delta_v^- = \{e-1\}$. For a fixed vertex v, we assume buffering queues at the beginning of each outgoing arc $e \in \delta_v^+$, and we define the following dynamics for a buffering queue $\mathcal{Q}^e$ of processor $e \in \delta_v^+$ and for $e - 1 \in \delta_v^-$:

$$\partial_t \mathcal{Q}^e(t,y) = \Phi^{e-1}(x_v^{e-1}, y, t) - \Phi^e(x_v^e, y, t), \tag{5.54a}$$

$$\Phi^e(x_v^e, y, t) := H(\mu^e - \beta^e(x_v^e, y, t)) v^e(x_v^e, y, t) f^e(x_v^e, y, t). \tag{5.54b}$$

For a given priority y, (5.54) resembles just the mass flux conservation through a vertex. The equation is underdetermined since for an outgoing arc $e \in \delta_v^+$ we can distribute the inflow among the queue $\partial_t \mathcal{Q}^e$ and processor given by $\Phi^e(x_v^e, y, t)$. In the case of a vertex

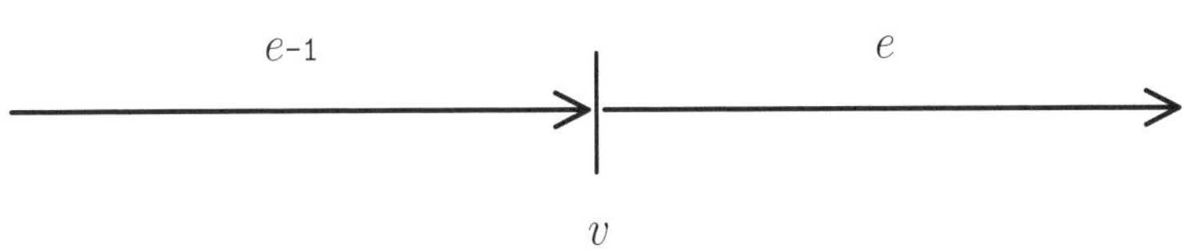

Figure 5.12. *The connection of two processors at a fixed vertex v.*

v of arbitrary degree, (5.54) has to be modified as follows: We assume the distribution of mass flux from incoming to outgoing processors to be known. Hence, a time- and attribute-dependent matrix $A^{v,e}(t,y) = (\alpha_{\bar{e},e})_{\bar{e},e}$, analogously to Definition 5.2 and (5.36). Then, (5.54) has to be replaced by

$$\partial_t \mathcal{Q}^e(t,y) = A^{v,e}(t,y) \sum_{\bar{e}\in\delta_v^-} \Phi^{\bar{e}}(x_v^{\bar{e}}, y, t) - \Phi^e(x_v^e, y, t). \tag{5.55}$$

Once we fix the value of Φ^e, we obtain the boundary value for the kinetic variable f^e by (5.54b). We propose the following consideration to fix the value of the outgoing flux Φ^e: We might have parts with different priorities reaching the vertex v on the incoming arc $e-1$. We assume that parts are passed through the vertex according to their priority. In particular, high priority parts $Y_1 < Y_2$ may be immediately passed to the next processor, whereas parts of lower priority Y_2 are buffered in the corresponding queue $\mathcal{Q}^e(Y_2,t)$. This corresponds to the governing equation for the dynamics inside each processor (5.53).

Additionally, at each vertex v, we introduce a function $t \to \mathcal{Y}^v(t)$ (called the pointer); see Figure 5.13. The pointer indicates the lowest priority part which is still processed on the outgoing arc e.

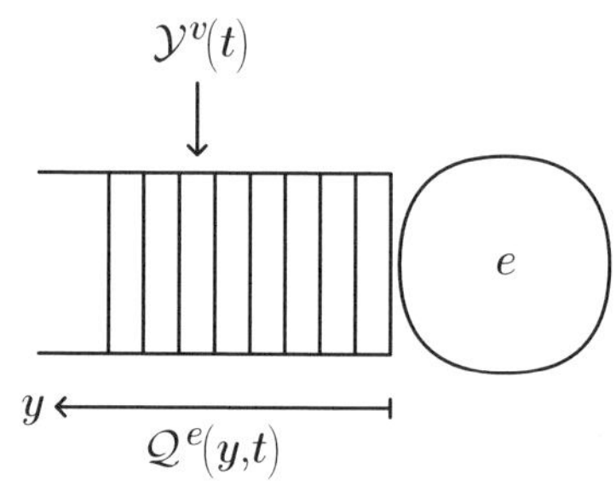

Figure 5.13. *The pointer variable $\mathcal{Y}^v(t)$ operates on the queue $\mathcal{Q}^e(y,t)$ of processor e.*

We derive an equation for the dynamics of $\mathcal{Y}^v(t)$ and expressions for Φ^e by considering a time-discrete situation first. Assume that at time t_n the pointer $\mathcal{Y}^v(t_n)$ is such that all of the incoming flux has been processed, i.e.,

$$\int_{-\infty}^{\mathcal{Y}^v(t_n)} \Phi^{e-1}(x_v^{e-1}, y, t_n)\, dy = \mu^e. \tag{5.56}$$

Then, at time t_{n+1}, we have to distinguish two cases: Either the inflow $\Phi^{e-1}(x_v^{e-1}, y, t_{n+1})$ of parts with priority $y < \mathcal{Y}^v(t_n)$ exceeds the maximal capacity μ^e of the outgoing processor or it does not. In the first case, the pointer will move towards higher priorities, and, in the second case, the pointer moves towards lower priorities and queues will start to empty. To be more precise, the total inflow at time $t_{n+1} = t_n + \Delta t$ of parts with priority less than $\mathcal{Y}^v(t_n)$ is given by

$$\varphi := \int_{-\infty}^{\mathcal{Y}^v(t_n)} \Phi^{e-1}(x_v^{e-1}, y, t_{n+1})\, dy. \tag{5.57}$$

Now, in the first case, $\varphi > \mu^e$ and $\mathcal{Y}^v(t_{n+1})$ decreases such that $\mathcal{Y}^v(t_{n+1})$ satisfies

$$\int_{-\infty}^{\mathcal{Y}^v(t_{n+1})} \Phi^{e-1}(x_v^{e-1}, y, t_{n+1})\, dy = \mu^e. \tag{5.58}$$

In the second case, $\varphi < \mu^e$ and the remaining capacity on processor e of $\mu^e - \varphi$ is filled up by parts of lower priority buffered in the queue or just arriving at the vertex. A sketch of the situation with two discrete pointer values is depicted in Figure 5.14.

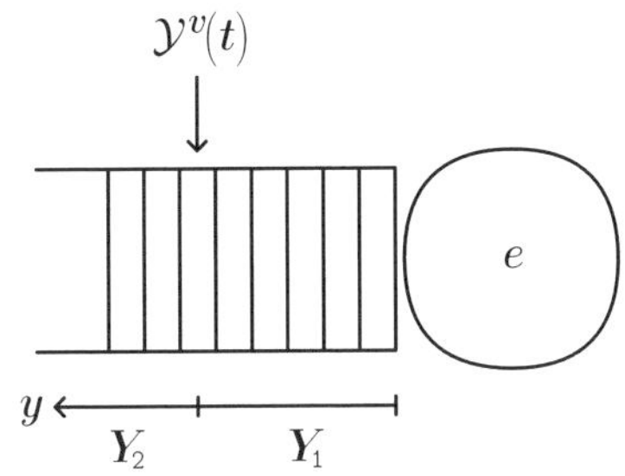

Figure 5.14. *In the case of $K = 2$ and the predefined relation $Y_1 < Y_2$, the pointer indicates high (labeled by Y_1) and low (labeled by Y_2) priority parts.*

The pointer $\mathcal{Y}$ increases such that $\mathcal{Y}^v(t_{n+1})$ satisfies

$$\int_{\mathcal{Y}^v(t_n)}^{\mathcal{Y}^v(t_{n+1})} \left(\Delta t \Phi^{e-1}(x_v^{e-1}, y, t_{n+1}) + \mathcal{Q}^e(t_{n+1}, y)\right) dy = (\mu^e - \varphi)\Delta t. \tag{5.59}$$

Formally, from (5.58), (5.59) for $\Delta t \to 0$ and regular Φ^{e-1}, we obtain an equation for $\mathcal{Y}^v(t)$:

$$\text{if } \frac{d\mathcal{Y}^v}{dt} < 0: \quad \int_{-\infty}^{\mathcal{Y}^v(t)} \Phi^{e-1}(x_v^{e-1}, y, t)\, dy - \mu^e = 0; \tag{5.60a}$$

$$\text{if } \frac{d\mathcal{Y}^v}{dt} > 0: \quad \mathcal{Q}^e(t, \mathcal{Y}^v(t))\frac{d\mathcal{Y}^v(t)}{dt} = \mu^e - \int_{-\infty}^{\mathcal{Y}^v(t)} \Phi^{e-1}(x_v^{e-1}, y, t)\, dy. \tag{5.60b}$$

In both cases, the outflow $\Phi^e(x_v^e, y, t)$ is determined according to the priority of the parts:

$$\begin{aligned} \Phi^e(x_v^e, y, t) &= \Phi^{e-1}(x_v^{e-1}, y, t) H(\mathcal{Y}^v(t) - y) \\ &+ \left(\mu^e - \int_{-\infty}^{\mathcal{Y}^v(t)} \Phi^{e-1}(x_v^{e-1}, y, t)\, dy\right) \delta(\mathcal{Y}^v(t) - y). \end{aligned} \tag{5.61}$$

In particular, (5.61) implies $\Phi^e(x_v^e, y, t) = 0$ for $y > \mathcal{Y}^v(t)$; i.e., no parts are processed with a lower priority than $\mathcal{Y}^v(t)$. If, nevertheless, such parts arrive, they will be buffered due to (5.54) and (5.61). If, on the other hand, we receive a part with priority $y < \mathcal{Y}^v(t)$, then it will be immediately passed to the outgoing arc, since in this case (5.61) yields $\Phi^e(x_v^e, y, t) = \Phi^{e-1}(x_v^{e-1}, y, t)$.

Moreover, in the case of multiple outflow arcs, we have a pointer variable $\mathcal{Y}^v(t) = \mathcal{Y}_e^v(t)$ for each outgoing arc $e \in \delta_v^+$. Furthermore, the definition (5.61) of the outgoing flux

Φ^e has to be replaced by (5.62):

$$\Phi^e(x_v^e, y, t) = A^{v,e}(t,y) \sum_{\bar{e}\in\delta_v^-} \Phi^{\bar{e}}(x_v^{\bar{e}}, y, t) H(\mathcal{Y}_e^v(t) - y)$$
$$+ \left(\mu^e - \int_{-\infty}^{\mathcal{Y}_e^v(t)} A^{v,e}(t,y) \sum_{\bar{e}\in\delta_v^-} \Phi^{\bar{e}}(x_v^{\bar{e}}, y, t) dy \right) \delta(\mathcal{Y}_e^v(t) - y). \tag{5.62}$$

Remark 5.5.2. *The well-posedness of the unsmooth dynamical problem* (5.60) *for the pointer variable* $\mathcal{Y}^v$ *is in general not obvious. However, considering "regularized" dynamics (see the remark below), well-posedness follows from standard considerations.*

We provide the following reformulation of (5.60) for determining $\mathcal{Y}^v(t)$:

$$\mathcal{Y}^v(t) := \min \left\{ \min\{\mathcal{Y} : \mathcal{Q}^e(t,\mathcal{Y}) \neq 0\}; \quad \mathcal{Y} : \int_{-\infty}^{\mathcal{Y}} \Phi^{e-1}(x_v^{e-1}, y, t)\, dy = \mu^e \right\}. \tag{5.63}$$

Equation (5.63) determines the pointer $\mathcal{Y}^v(t)$ by comparing the priorities of parts arriving and the priority of those which are already in a queue. Equations (5.61) and (5.54) remain unchanged. Obviously, we recover the first case of (5.60). To recover the second case of (5.60) we integrate (5.54) on $(-\infty, \mathcal{Y}^v(t))$ and replace Φ^e by its definition (5.61):

$$\int_{-\infty}^{\mathcal{Y}^v(t)-0} \partial_t \mathcal{Q}^e(y,t)\, dy = -\left(\mu^e - \int_{-\infty}^{\mathcal{Y}^v(t)+0} \Phi^{e-1}(x_v^{e-1}, y, t)\, dy \right).$$

Due to (5.63) we can assume that $\mathcal{Q}^e(y,t) = 0$ for $y < \mathcal{Y}^v(t)$, and therefore

$$\int_{-\infty}^{\mathcal{Y}^v(t)-0} \partial_t \mathcal{Q}^e(t,y)\, dy = \int_{-\infty}^{\infty} \partial_t \mathcal{Q}^e(t,y) H(\mathcal{Y}^v(t) - y)\, dy$$
$$= \partial_t \left(\int_{-\infty}^{\infty} \mathcal{Q}^e(t,y) H(\mathcal{Y}^v(t) - y)\, dy \right) - \int_{-\infty}^{\infty} \delta(\mathcal{Y}^v(t) - y) \frac{d\mathcal{Y}^v(t)}{dt} \mathcal{Q}^e(t,y)\, dy$$
$$= 0 - \mathcal{Q}^e(t, \mathcal{Y}^v(t)) \frac{d\mathcal{Y}^v(t)}{dt},$$

which finally yields the second case of (5.60). Those computations are only formal but illustrate the meaning of the pointer $\mathcal{Y}^v$. Moreover, (5.63) will be used to obtain numerical results.

Remark 5.5.3. *Due to* (5.61)*, the solution* Φ^e *may contain delta distributions. To remedy this unsatisfactory fact, we introduce a smoothed version of the dynamics given above. If* $\Phi^{e-1}(x_v^{e-1}, y, t)$ *has a delta concentration at* $\mathcal{Y}^v(t)$*, then the derivation of the continuous equations has to be modified and* (5.60) *is replaced by the following:*

$$\text{if } \frac{d\mathcal{Y}^v}{dt} < 0: \quad \int_{-\infty}^{\mathcal{Y}^v(t)-0} \Phi^{e-1}(x_v^{e-1}, y, t)\, dy - \mu^e = 0;$$
$$\text{if } \frac{d\mathcal{Y}^v}{dt} > 0: \quad \mathcal{Q}^e(t, \mathcal{Y}^v(t)) \frac{d\mathcal{Y}^v(t)}{dt} = \mu^e - \int_{-\infty}^{\mathcal{Y}^v(t)+0} \Phi^{e-1}(x_v^{e-1}, y, t)\, dy.$$

Therein, $\mathcal{Y}^v(t) \pm 0$ denotes the value of the pointer after and before the delta concentration in Φ^{e-1}.

To motivate the smoothed version we first consider the discrete case. We assume a discrete and ordered number of properties $y \in \{\mathcal{Y}_k, k = 0, 1, \dots\}$. Let K be determined such that the discrete form of (5.63) holds, i.e.,

$$K := \min\left\{\min\{k : \mathcal{Q}^e(t,\mathcal{Y}_k) \neq 0\};\ \tilde{k} : \left\{\sum_{j=0}^{\tilde{k}-1} \Phi^{e-1}(t,\mathcal{Y}_j) \leq \mu^e < \sum_{j=0}^{\tilde{k}} \Phi^{e-1}(t,\mathcal{Y}_j)\right\}\right\}. \tag{5.64}$$

Then the discrete dynamics are, for $k = 0, 1, \dots$,

$$\text{if } k < K: \quad \Phi^e(t,\mathcal{Y}_k) = \Phi^{e-1}(\mathcal{Y}_k, t); \tag{5.65a}$$

$$\text{if } k = K: \quad \Phi^e(t,\mathcal{Y}_k) = \mu^e - \sum_{j=0}^{K-1} \Phi^{e-1}(t,\mathcal{Y}_j); \tag{5.65b}$$

$$\text{if } k > K: \quad \Phi^e(t,\mathcal{Y}_k) = 0; \tag{5.65c}$$

and

$$\partial_t \mathcal{Q}^e(t,\mathcal{Y}_k) = \Phi^{e-1}(t,\mathcal{Y}_k) - \Phi^e(t,\mathcal{Y}_k) \quad \forall k = 0, 1, \dots. \tag{5.66}$$

In the case $k = K$ we have a remaining capacity of $\mu^e - \sum_{j=0}^{K-1} \Phi^{e-1}(t,\mathcal{Y}_j)$ on the outgoing arc which we let pass. Now, depending on the two possibilities in (5.64), this setting implies the following: If $\sum_{j=0}^{K} \Phi^{e-1}(t,\mathcal{Y}_j) > \mu^e$, we then obtain $\partial_t \mathcal{Q}^e(t,\mathcal{Y}_K) > 0$ and we store the additional parts in the buffering queue. If, on the other hand, $\sum_{j=0}^{K} \Phi^{e-1}(t,\mathcal{Y}_j) < \mu^e$, then $\partial_t \mathcal{Q}^e(t,\mathcal{Y}_K) < 0$ and we reduce the queue $\mathcal{Q}^e_K$. This is possible, since due to (5.64) $\mathcal{Q}^e(t,\mathcal{Y}_K) > 0$.

A continuous reformulation of (5.65) is given by

$$\bar{\Phi}^e(t,y) = \int_{-\infty}^{y} \Phi^{e-1}(x_v^{e-1},\eta,t) H(\mathcal{Y}^v - \eta) + \left(\mu^e - \int_{-\infty}^{\mathcal{Y}^v} \Phi^{e-1}(x_v^{e-1},s,t)ds\right)\delta(\mathcal{Y}^v - \eta)d\eta$$

for $\mathcal{Y}^v$ arbitrary and $\Phi^e(t,y) = \partial_y \bar{\Phi}^e(t,y)$. Further, if $\mathcal{Y}^v = \mathcal{Y}^v(t) < \infty$, then $\bar{\Phi}^e(t,y)$ is discontinuous at $y = \mathcal{Y}^v(t)$. Hence, we smooth $\bar{\Phi}^e(t,y)$ such that

$$\bar{\Phi}^{e,\epsilon}(t,y) = \bar{\Phi}^e(t,y), \quad y \leq \mathcal{Y}^v(t),$$
$$\bar{\Phi}^{e,\epsilon}(t,y) = \bar{\Phi}^e(\mathcal{Y}^v(t),t) + \frac{1}{\epsilon}\left(\mu^e - \bar{\Phi}^e(\mathcal{Y}^v(t),t)\right)(y - \mathcal{Y}^v(t)), \quad \mathcal{Y}^v(t) < y < \mathcal{Y}^v(t) + \epsilon,$$
$$\bar{\Phi}^{e,\epsilon}(t,y) = \mu^e, \quad y \geq \mathcal{Y}^v(t) + \epsilon,$$

holds, and we define the smoothed dynamics as $\Phi^{e,\epsilon}(t,y) = \partial_y \bar{\Phi}^{e,\epsilon}(t,y)$ and $\partial_t \mathcal{Q}^e(t,y) = \Phi^e(t,y) - \bar{\Phi}^{e,\epsilon}(t,y)$. In particular, we still obtain $\partial_t \mathcal{Q}^e(t,y) = 0$ for $y < \mathcal{Y}^v(t)$.

We derive equations for the moments on each arc. In the case of the network model, these moment equations have to be accompanied by suitable boundary conditions. We derive boundary conditions from the equations for the buffering queues.

On each arc $e \in \mathcal{A}$ we introduce the moments m_j^e and moment fluxes F_j^e for $j = 0, \dots, J-1$ similarly to (5.46) for $x \in [a^e, b^e]$, $t > 0$:

$$m_j^e(t,x) = \int y^j f^e(x,y,t) dy, \tag{5.67a}$$

$$F_j^e(t,x) = \int y^j \left(H(\mu^e - \beta^e(x,y,t)) v^e(x,y,t) f^e(x,y,t) \right) dy. \tag{5.67b}$$

Further, we use on each arc $e \in \mathcal{A}$ the same equilibrium closure as for the single supply chain, namely,

$$f_e^\epsilon(x,y,t) = \sum_{k=1}^{K} \rho_k^e(t,x) \frac{1}{\epsilon} \phi' \left(\frac{y - Y_k(t,x)}{\epsilon} \right). \tag{5.68}$$

Of course, we obtain an analogous result as in Proposition 5.10: The dynamic on each arc e of a multiphase supply chain network is governed by a set of equations (5.69) if the phases Y_k, $k = 1, \dots, K$, are mutually distinct:

$$\partial_t \rho_k^e + \partial_x q_k^e = 0, \tag{5.69a}$$

$$\partial_t (\rho_k^e Y_k) + \partial_x (q_k^e Y_k) = 0, \tag{5.69b}$$

$$q_k^e = \rho_k^e v_k^e Z_k^e, \tag{5.69c}$$

where ρ^e and q^e are the density and flux on arc e, $v_k^e = v^e(x, Y_k, t)$, and Z_k^e is defined analogously to Proposition 5.10 and with

$$m_j^e(t,x) = \sum_{k=1}^{K} \rho_k^e(t,x) Y_k^j(t,x) \quad \text{and} \quad F_j^e(t,x) = \sum_{k=1}^{K} q_k^e(t,x) Y_k^j(t,x) \tag{5.70}$$

in the limit $\epsilon \to 0$. In a next step, we derive appropriate expressions for the boundary conditions for multiphase flow and we consider a fixed vertex $v \in \mathcal{V}$ with incoming arc $e-1$ and outgoing arc e. As noted in [3], we need $J = 2K$ moments for a K-phase flow model to balance the number of $2K$ unknowns (ρ_k^e, Y_k) on each arc e. Here, we have to prescribe boundary conditions in terms of the K fluxes q_k^e and $q_k^e Y_k$ at $x = a^e$. Formally, these conditions are given after integrating (5.61) against y^j for $j = 0, \dots, 2K$:

$$F_j^e(x_v^e, t) = \int_{-\infty}^{\infty} y^j \Phi^e(x_v^e, y, t)\, dy = \int_{-\infty}^{\mathcal{Y}^v(t)} y^j \Phi^{e-1}(x_v^{e-1}, y, t)\, dy \tag{5.71}$$
$$+ \int_{-\infty}^{\infty} y^j \left(\mu^e - \int_{-\infty}^{\mathcal{Y}^v(t)} \Phi^{e-1}(x_v^{e-1}, \eta, t)\, d\eta \right) \delta(\mathcal{Y}^v(t) - y)\, dy.$$

Using the closure relation (5.68) we derive explicit conditions for the multiphase model of K phases: We assume the K phases $Y_1, \dots, Y_K$ to be mutually distinct and ordered $Y_{k-1} < Y_k$. Further, we consider the reformulation (5.63) of the pointer equation for $\mathcal{Y}^v(t)$. Since we treat a finite number of phases, (5.63) implies that for each $t > 0$

$$\mathcal{Y}^v(t) \in \{Y_1, \dots, Y_K, +\infty\}. \tag{5.72}$$

For the movement of the pointer $\mathcal{Y}^v(t)$ we discuss some cases of interest:

1. If $\mathcal{Y}^v(t) = +\infty$, then (5.71) implies

$$F_j^e(x_v^e, t) = F_j^{e-1}(x_v^{e-1}, t) \quad \forall j = 0, \dots, J-1. \tag{5.73}$$

In this case there is enough capacity on processor e, so that parts of any priority are processed. To obtain boundary conditions for (5.69), we view (5.73) as a system of J equations in the variables $(q_k^e, q_k^e Y_k)$, $k = 1, \dots, K$; i.e., (5.73) is equivalent to

$$\sum_{k=1}^{K} q_k^e(x_v^e, t) Y_k^j(x_v^e, t) = \sum_{k=1}^{K} q_k^{e-1}(x_v^{e-1}, t) Y_k^j(x_v^{e-1}, t) \quad \forall j = 0, \dots, J-1.$$

Obviously, a solution $(q_k^e, q_k^e Y_k^e)$ to the previous set of equations is given by

$$Y_k(x_v^e, t) = Y_k(x_v^{e-1}, t), \quad k = 1, \dots, K, \tag{5.74a}$$

$$q_k^e(x_v^e, t) = q_k^{e-1}(x_v^{e-1}, t), \quad k = 1, \dots, K. \tag{5.74b}$$

Now, (5.74) defines the boundary conditions for (5.69) on the outgoing arc e.

2. If $\mathcal{Y}^v(t) = Y_1, Y_1 = Y_1(x_v^{e-1}, t)$, then (5.71) yields

$$\begin{aligned} F_j^e(x_v^e, t) &= q_1^{e-1}(x_v^{e-1}, t) Y_1^j + \int_{-\infty}^{\infty} y^j \left(\mu^e - q_1^{e-1}(x_v^{e-1}, t)\right) \delta(Y_1 - y)\, dy + O(\epsilon) \\ &= \mu^e Y_1^j + O(\epsilon) \quad \forall j, \end{aligned}$$

and, as before, we derive the following boundary conditions for $(q_k^e, q_k^e Y_k^e)$:

$$\begin{aligned} &Y_k(x_v^e, t) = Y_k(x_v^{e-1}, t), \quad k = 1, \dots, K, \\ &q_1^e(x_v^e, t) = \mu^e, \\ &q_k^e(x_v^e, t) = 0, \quad k = 2, \dots, K. \end{aligned}$$

3. If $\mathcal{Y}^v(t) = Y_\kappa, \kappa > 1$, $Y_\kappa = Y_\kappa(x_v^{e-1}, t)$, then (5.71) yields

$$\begin{aligned} F_j^e(x_v^e, t) &= \sum_{k=1}^{\kappa} q_k^{e-1}(x_v^{e-1}, t) Y_k^j \\ &\quad + \int_{-\infty}^{\infty} y^j \left(\mu^e - \int_{-\infty}^{Y_\kappa} \Phi^{e-1}(x_v^{e-1}, \eta, t)\, d\eta\right) \delta(Y_\kappa - y)\, dy + O(\epsilon) \\ &= \sum_{k=1}^{\kappa} q_k^{e-1}(x_v^{e-1}, t) Y_k^j + Y_\kappa^j \left(\mu^e - \sum_{k=1}^{\kappa} q_k^{e-1}(x_v^{e-1}, t)\right) + O(\epsilon) \\ &= \sum_{k=1}^{\kappa-1} q_k^{e-1}(x_v^{e-1}, t) Y_k^j + \left(\mu^e - \sum_{k=1}^{\kappa-1} q_k^{e-1}(x_v^{e-1}, t)\right) Y_\kappa^j + O(\epsilon) \quad \forall j, \end{aligned}$$

and therefore we derive the following conditions on $(q_k^e, q_k^e Y_k^e)$:

$$Y_k(x_v^e,t) = Y_k(x_v^{e-1},t), \quad k=1,\dots,K, \tag{5.75a}$$

$$q_k^e(x_v^e,t) = q_k^{e-1}(x_v^{e-1},t), \quad k=1,\dots,\kappa-1, \tag{5.75b}$$

$$q_\kappa^e(x_v^e,t) = \mu^e - \sum_{k=1}^{\kappa-1} q_k^{e-1}(x_v^{e-1},t). \tag{5.75c}$$

In this scenario, all parts with higher priority than κ are processed and the remaining capacity of processor e will be filled with parts of priority κ.

Next, we derive the equations for the buffering queues with the assumption (5.68). We integrate (5.54a) also against y^j for $j=0,\dots,J=2K$, and we obtain

$$\int_{-\infty}^{\infty} y^j \partial_t \mathcal{Q}^e(t,y)dy = F_j^{e-1}(x_v^{e-1},t) - F_j^e(x_v^e,t), \quad j=0,\dots,J-1. \tag{5.76}$$

Further, considering (5.70) and the equality $Y_k(x_v^e,t) = Y_k(x_v^{e-1},t)$ derived before, we obtain

$$\int_{-\infty}^{\infty} y^j \partial_t \mathcal{Q}^e(t,y)dy = \sum_{k=1}^{K} \left(q_k^{e-1}(x_v^{e-1},t) - q_k^e(x_v^e,t) \right) Y_k^j(x_v^e,t), \quad j=0,\dots,J-1. \tag{5.77}$$

Equality (5.77) can be satisfied by introducing a finite number of queues $\pi_k^e(t)$, $k=1,\dots,K$, corresponding to the finite number of priorities. Using an approach similar to the one for the equilibrium closure (5.68) and introducing the smoothed sum of delta concentrations at Y_k, we obtain

$$\partial_t \mathcal{Q}^{e,\epsilon}(t,y) = \sum_{k=1}^{K} \frac{1}{\epsilon}\phi'\left(\frac{y-Y_k(t,x)}{\epsilon}\right)\frac{d}{dt}\pi_k^e(t). \tag{5.78}$$

Using the above ansatz, (5.77) reduces in the limit $\epsilon \to 0$ to an equation for the queues π_k^e, that is, a queue for priority Y_k,

$$\frac{d}{dt}\pi_k^e(t) = q_k^{e-1}(x_v^{e-1},t) - q_k^e(x_v^e,t), \quad k=1,\dots,K. \tag{5.79}$$

Summarizing, the macroscopic supply chain model in network formulation is given by (5.69), (5.75), and (5.79).

Remark 5.5.4. *The derivation of the macroscopic equations can be adapted to general network topologies in the following way: We consider the case of a vertex $v \in \mathcal{V}$ of arbitrary degree and recall the assumptions made for deriving* (5.55) *and* (5.62)*: There exists a matrix $A^{v,e}(t,y)$ controlling the distribution of incoming parts among the outgoing processors. The reformulation of the equation for each pointer $\mathcal{Y}_e^v(t)$ is now given by*

$$\mathcal{Y}_e^v(t) := \min\left\{ \min\{\mathcal{Y} : \mathcal{Q}^e(\mathcal{Y},t) \neq 0\};\ \mathcal{Y} : \int_{-\infty}^{\mathcal{Y}} A^{v,e}(t,y) \sum_{\tilde{e}\in\delta_v^-} \Phi^{\tilde{e}}(x_v^{\tilde{e}},y,t)dy = \mu^e \right\}.$$

Now, fix an outgoing arc $e \in \delta_v^+$. Then, we first average on all incoming arcs $\tilde{e} \in \delta_v^-$ with weights given by $A^{v,e}(t,y)$ and obtain the moments $\bar{F}_j$ incoming to arc e by

$$\bar{F}_j(t) = A^{v,e}(t,y) \sum_{\tilde{e}\in\delta_v^-} \int_{-\infty}^{\infty} y^j\, \Phi^{\tilde{e}}(x_v^{\tilde{e}}, y, t)\, dy$$

$$= \sum_{k=1}^{K} A^{v,e}(t, Y_k^e) \sum_{\tilde{e}\in\delta_v^-} (t, Y_k^{\tilde{e}})(x_v^{\tilde{e}}, t) q_k^{\tilde{e}}(x_v^{\tilde{e}}, t) \left(Y_k^{\tilde{e}}\right)^j (x_v^{\tilde{e}}, t)$$

for $j = 0,\dots,J-1$. As in [3] *we decompose $F_j(t)$ in $\bar{\mathcal{Y}}_k$ and $\bar{q}_k$ such that*

$$\bar{F}_j(t) = \sum_{k=1}^{K} \bar{\mathcal{Y}}_k^j(t)\, \bar{q}_k(t), \quad j = 0,\dots,J-1. \tag{5.80}$$

Finally, we proceed analogously to the previous discussion to determine $q_k^e(x_v^e, t)$, $\mathcal{Y}_k^e(x_v^e, t)$, where now the moment fluxes $\bar{q}_k$ and $\bar{\mathcal{Y}}_k$ replace the fluxes q_k^{e-1} and $\mathcal{Y}_k^{e-1}$, respectively. In particular, the case distinction discussed on page 79 *remains the same.*

At last, we specify the equations for the single-phase and the two-phase flows in the case of a linear network. We add further models resultant from (5.69), (5.75), and (5.79) for special choices of the processing velocities.

Single-Phase Flow

We conclude the derivation with the example of the single-phase flow equations on a network. The dynamics of the density of parts ρ^e for an attribute $Y := Y_1$ on arc e is then given by (5.69), i.e,

$$\partial_t \rho^e + \partial_x q^e = 0, \tag{5.81a}$$

$$\partial_t(\rho^e Y) + \partial_x(q^e Y) = 0, \tag{5.81b}$$

$$q^e = \min\{\mu^e, v^e(x,Y)\rho^e\}, \tag{5.81c}$$

and at a vertex $v \in \mathcal{V}$ the following coupling conditions between the arcs $e-1$ and e hold. According to (5.63), we have to distinguish the cases:

$$\text{if } \pi^e(t) = 0: \quad q^e(x_v^e, t) = \min\{q^{e-1}(x_v^{e-1}, t), \mu^e\}; \tag{5.82a}$$

$$\text{if } \pi^e(t) \neq 0: \quad q^e(x_v^e, t) = \mu^e. \tag{5.82b}$$

Moreover, $\partial_t \mathcal{Q}^e(t,y) = \delta(y-Y)\frac{d}{dt}\pi^e(t)$ and $Y(x_v^e, t) = Y(x_v^{e-1}, t)$.

Two-Phase Flow

In the case of the two-phase flow we assume $Y_1 < Y_2$. The dynamics on an arc e is given by, for $k = 1,2$,

$$\partial_t \rho_k^e + \partial_x q_k^e = 0, \tag{5.83a}$$

$$\partial_t(\rho_k^e Y_k) + \partial_x(q_k^e Y_k) = 0, \tag{5.83b}$$

where q_k^e is given by Remark 5.5.1. The dynamics of the queues $\pi_k^e(t)$, $k = 1,2$, is given by (5.79) and $\mathcal{Q}^e(y,t) = \delta(y - Y_1)d_t\pi_1(t) + \delta(y - Y_2)d_t\pi_2(t)$. For the definition of the outgoing fluxes q_k^e, $k = 1,2$, the previous case distinction on page 79 can be rewritten in the following condensed form:

$$\begin{aligned}
q_1^e(x_v^e,t) &= q_1^{e-1}(x_v^{e-1},t) + \delta(Y_1 - \mathcal{Y}^v(t))((\mu^e - q_1^{e-1}(x_v^{e-1},t)),\\
q_2^e(x_v^e,t) &= q_2^{e-1}(x_v^{e-1},t) - \delta(Y_1 - \mathcal{Y}^v(t)q_2^{e-1}(x_v^{e-1},t)\\
&\quad + \delta(Y_2 - \mathcal{Y}^v(t))((\mu^e - q_2^{e-1}(x_v^{e-1},t) - q_1^{e-1}(x_v^{e-1},t)),\\
Y_k(x_v^e,t) &= Y_k(x_v^{e-1},t).
\end{aligned}$$

Moreover, the determination (5.63) of $\mathcal{Y}^v$ yields more complex formulas compared with (5.82), since we have a true dynamic of the pointer $\mathcal{Y}^v$. We distinguish several cases (where the roles must be exchanged for $Y_2 < Y_1$):

$$\mathcal{Y}^v(t) = \begin{pmatrix} Y_1, & \text{if } \pi_1^e(t) \neq 0 \text{ or if } q_1^{e-1}(x_v^{e-1},t) > \mu^e \\ Y_2, & \text{if } \pi_1^e(t) = 0,\ \pi_2^e(t) \neq 0 \text{ or if } q_1^{e-1}(x_v^{e-1},t) + q_2^{e-1}(x_v^{e-1},t) \geq \mu^e > q_1^{e-1}(x_v^{e-1},t) \\ +\infty, & \text{if } \pi_1^e(t) = \pi_2^e(t) = 0 \text{ and if } q_1^{e-1}(x_v^{e-1},t) + q_2^{e-1}(x_v^{e-1},t) \leq \mu^e \end{pmatrix}.$$

This finishes the two-phase case. In the two-phase model we may illustrate the dynamics by comparing with Figure 5.15. We consider a linear chain consisting of three processors with different processing rates μ. High priority parts Y_1 are depicted in gray and low priority parts Y_2 in black. In a fixed unit time step, processor 1 is able to process all ingoing parts due to the sufficiently large capacity $\mu = 4$. In this case, the pointer $\mathcal{Y}^v(t)$ will be infinity. However, the second processor has a decreasing maximal capacity, and the lower priority parts must be stored while the higher ones are in progress. Now the pointer moves to Y_2. The last processor reflects the situation where the capacity is insufficient to handle the incoming parts and as a consequence the queue is filled, i.e., $\mathcal{Y}^v(t) = Y_1$.

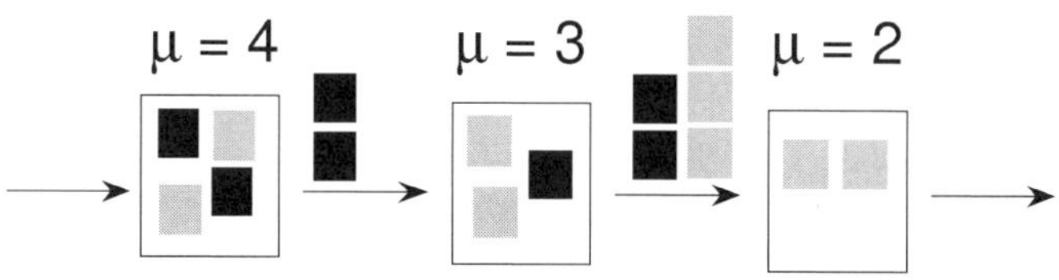

Figure 5.15. *Supply chain governed by a two-phase model.*

Deduced Models

Finally, we present two models derived from the multiphase network model for a special choice of the processing velocity v^e.

1. Equations (5.82), (5.50) are obtained by assuming a constant velocity $v^e(x,Y) = v^e$. Only in this case (5.81c) simplifies and we obtain

$$\partial_t \rho^e + \partial_x v^e \rho^e = 0, \tag{5.84}$$

$$\partial_t(\rho^e Y) + \partial_x(v^e \rho^e Y) = 0, \tag{5.85}$$

and due to (5.75) we ensure that $q^e = v^e\rho^e \leq \mu^e$.

2. We have no restriction on arbitrary velocity relations $v^e(x,Y)$. For the particular choice

$$v^e(x,Y) = Y, \tag{5.86}$$

we obtain a flux-restricted pressureless gas law for the evolution for the transport of the property Y and the particle density ρ^e:

$$\partial_t \rho^e + \partial_x \min\{\mu^e, Y\rho^e\} = 0, \tag{5.87a}$$

$$\partial_t Y + \min\{\mu^e/\rho^e, Y\}\partial_x Y = 0. \tag{5.87b}$$

5.6 Summary

In this chapter, we have derived a model for production networks by extending the work of Armbruster, Degond, and Ringhofer; cf. Chapter 3. The main advantage of the new model is the easy adaption to complex network topologies by defining so-called coupling conditions at vertices. In comparison with other existing models, the presented network model consists of queues and processors given by a coupled system of partial and ordinary differential equations. From the point of view of application areas further advantages are the following: The partial differential equation approach gives a guideline to develop new and dynamically accurate models. And, the approach opens a way to introduce nonlinearities in a straightforward and consistent way and to treat nonlinear problems like chemical production by appropriate methods. It is even possible to show the existence of a solution in a mathematical sense since the network equations do not allow for delta distributions which complicate theoretical treatment. As an extension of the basic model, we have introduced a modified supply network model containing policy attributes where again the concept of installing queues is used for the coupling of different processors.

Chapter 6

Continuum-Discrete Models

In this chapter we introduce a model which extends the one of Armbruster, Degond, and Ringhofer presented in Section 3.3 of Chapter 3. More precisely, we want to introduce a dynamics for the processing rate; hence each arc is modelled by a system of two equations: a conservation law for the good density ρ, and an evolution equation for the processing rate μ.

Once we introduce the model, we discuss possible choices of solutions at nodes guaranteeing the conservation of fluxes. Keeping the analogy to Riemann problems, we call the latter Riemann Solver at nodes. Our first choice is to fix the rule:

SC1 The incoming density flux is equal to the outgoing density flux. Then, if a solution with only waves in the density ρ exists, then such solution is taken; otherwise the minimal μ wave is produced.

Rule SC1 corresponds to the case in which processing rate adjustments are done only if necessary, while the density ρ can be regulated more freely. Thus, it is justified in all situations in which processing rate adjustments require rebuilding of the supply chain, while density adjustments are operated easily (e.g., by stocking). Even if rule SC1 is the most natural also from a geometric point of view, in the space of Riemann data, it produces waves only to lower the value of μ. As a consequence in some cases the value of the processing rate does not increase and it is not possible to maximize the flux. In order to avoid this problem we analyze two additional rules to solve dynamics at a node:

SC2 The objects are processed in order to maximize the flux with the minimal value of the processing rate.

SC3 The objects are processed in order to maximize the flux. Then, if a solution with only waves in the density ρ exists, then such a solution is taken; otherwise the minimal μ wave is produced.

The continuum-discrete model, regarding sequential supply chains, has been generalized to supply networks which consist of arcs and two types of nodes: nodes with one incoming arc and more outgoing ones and nodes with more incoming arcs and one outgoing arc.

The Riemann problems are solved fixing two "routing" algorithms:

RA1 Goods from an incoming arc are sent to outgoing ones according to their final destination in order to maximize the flux over incoming arcs. Goods are processed, ordered by arrival time (FIFO policy).

RA2 Goods are processed by arrival time (FIFO policy) and are sent to outgoing arcs in order to maximize the flux over incoming and outgoing arcs.

For both routing algorithms the flux of goods is maximized considering one of the two additional rules, SC2 and SC3.

The chapter is organized as follows. Real examples are discussed in Section 6.1. Section 6.2 gives the basic definitions of supply chain and Riemann Solver. Then the dynamics inside an arc is studied in Section 6.3, providing also flux total variation estimates. In Section 6.4, general Riemann Solvers at junctions are discussed. Then in Section 6.5, a particular Riemann Solver according to rule SC1 is defined and explicit unique solutions are given. Section 6.6 provides the needed estimates for constructing solutions to Cauchy problems, and an example of unbounded increase of $T.V.(\rho)$ is shown. In Sections 6.7 and 6.8, new Riemann Solvers according to rules SC2 and SC3 are described. Section 6.9 provides analysis of equilibria in a node for Riemann Solvers which respect rules SC1, SC2, and SC3. The last section extends the model to simple supply networks.

6.1 Motivation

In this section we give some examples of real supply networks which can motivate the introduction of the model.

Let us analyze a supply network for assembling wine bottles, described in Figure 6.1(left). Bottles coming from arc e^1 are sterilized in node v^1. Then, the sterilized bottles with a certain probability α are directed to node v^3, where white wine is bottled, and with probability $1-\alpha$ to node v^4, where red wine is bottled. In nodes v^5 and v^6, bottles are labeled, respectively, for red and white wine. Finally, in node v^7, produced bottles are corked. Assume that red and white wine bottles are produced using two different bottle shapes. The bottles are addressed from arc e^2 to the outgoing arcs e^3 and e^4 in which they are filled up with white or red wine according to the bottle shape and thus according to the final destination: production of white or red wine bottles. In a model able to describe this

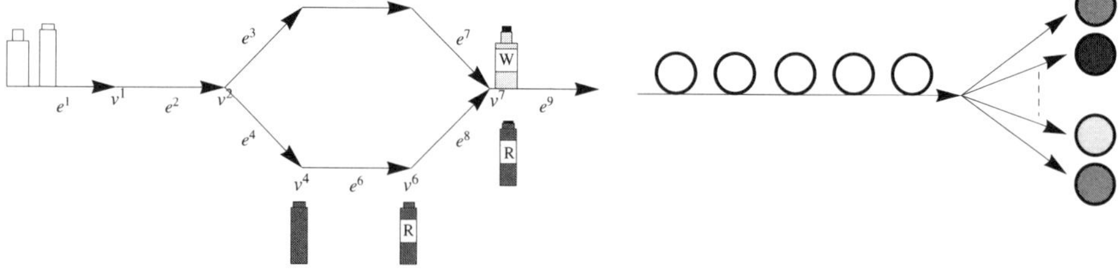

Figure 6.1. *Wine production network.*

situation, the dynamics at the node v^2 is solved using the RA1 algorithm. In fact it is not possible to redirect the bottles in order to maximize the production on both incoming and outgoing arcs, since bottles with white and red wine have different shapes.

A supply network of beach ball production is considered in Figure 6.1(right). The white balls are addressed towards n arcs in which they are colored using different colors. Since the aim of the factory is to maximize ball production independently from the colors, a mechanism is realized which addresses the balls on the outgoing arcs taking into account their loads in such a way to maximize flux on both incoming and outgoing arcs. It follows that a model realized to capture the behavior of the described supply network is based on algorithm RA2.

Let us now analyze an existing supply network where both algorithms show up naturally: the one for chip production of the San Carlo enterprise. The productive processes follow various steps that can be summarized as follows: When potatoes arrive at the enterprise, they are subjected to a goodness test. After this test, everything is ready for chip production, starting with potato washing in drinking water. After washing potatoes, they are skinned off, rewashed, and subjected to a qualification test. Then, they are cut in thin strips by an automatic machine and, finally, washed and dried by an air blow. At this point, potatoes are ready to be fried in vegetable oil for some minutes, and, after this, the surplus oil is dripped. Potatoes are then salted by a dispenser that nebulizes salt spreading it on potatoes. An opportune chooser is useful to select the best products. The final phase of the process is given by the potato confection. A simplified vision of the supply chain network is in Figure 6.2(top). In phases 1, 5, and 10 a discrimination is made in production in order to distinguish good and bad products. In such a sense, we can say that there is a statistical percentage α of product that follows the production steps, while the percentage $1-\alpha$ is the product discarded (obviously, the percentage α can be different for different

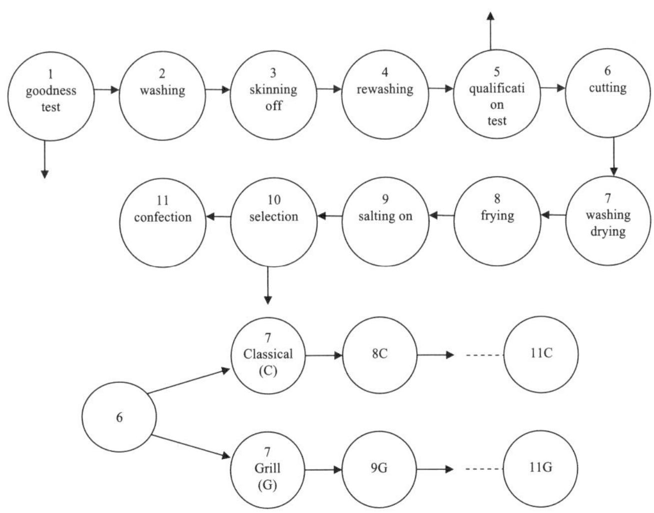

Figure 6.2. *Graph of the supply network for chip production (top) and possible arcs (bottom).*

phases). Therefore, the goods routing in these nodes follow algorithm RA1. On the other side, phase 6 concerns the potatoes cut: as the enterprise produces different types of fried potatoes (classical, grill, light, stick, etc.), different ways of cutting potatoes must be considered. Assume that, for simplicity, there are only two types of potato production; then the supply network is as in Figure 6.2(bottom). If the aim is only the production maximization independently from the type, then the potatoes are addressed from node 6 towards the outgoing arcs according to the RA2 algorithm.

6.2 Basic Definitions

We start from the conservation law model:

$$\rho_t + (\min\{\mu(t,x),\rho\})_x = 0. \tag{6.1}$$

To avoid problems of existence of solutions, we assume μ piecewise constant and an evolution equation of semilinear type:

$$\mu_t + \overline{V}\mu_x = 0, \tag{6.2}$$

where $\overline{V}$ is some constant velocity. Taking $\overline{V} = 0$, we may have no solution to a Riemann problem for the system (6.1)–(6.2) with data (ρ_l,μ_l) and (ρ_r,μ_r) if $\min\{\mu_l,\rho_l\} > \mu_r$. Since we expect the chain to influence backward the processing rate, we assume $\overline{V} < 0$ and for simplicity we set $\overline{V} = -1$.

We define a mixed continuum-discrete model in the following way. A supply chain consists of sequential processors, and it is again modelled by a real line seen as a sequence of arcs corresponding to intervals $[a^e,b^e]$ such that $[a^e,b^e]\cap\left[a^{e+1},b^{e+1}\right] = v^e$: a vertex separating arcs.

On each arc, the evolution is given by (6.1)–(6.2). On the other side, the evolution at nodes v^e is given solving Riemann problems for the density equation (6.1) with μ's as parameters. Such Riemann problems may still admit no solution as before if we keep the values of the parameters μ's constant; thus we expect μ waves to be generated and then follow (6.2). The vanishing of the characteristic velocity for (6.1), in case $\rho > \mu$, can provoke resonances with the nodes (which can be thought of as waves with zero velocities). Therefore, we slightly modify the model as follows.

Each arc e is characterized by a maximum density $\rho^e_{\max}$, a maximum processing rate $\mu^e_{\max}$, and a flux f^e_ε. For a fixed $\varepsilon > 0$, the dynamics is given by

$$\begin{cases} \rho_t + f^e_\varepsilon(\rho,\mu)_x = 0, \\ \mu_t - \mu_x = 0. \end{cases} \tag{6.3}$$

The flux is defined as

$$\text{(F)}\quad f^e_\varepsilon(\rho,\mu) = \begin{cases} \rho, & 0 \le \rho \le \mu, \\ \mu + \varepsilon(\rho-\mu), & \mu \le \rho \le \rho^e_{\max}, \end{cases}$$

$$f^e_\varepsilon(\rho,\mu) = \begin{cases} \varepsilon\rho + (1-\varepsilon)\mu, & 0 \le \mu \le \rho, \\ \rho, & \rho \le \mu \le \mu^e_{\max}; \end{cases}$$

see Figure 6.3.

The conservation law for the good density in (6.3) is a ε perturbation of (6.1) in the sense that $\|f - f_\varepsilon\|_\infty \leq C\varepsilon$, where f is the flux of (6.1). The equation has the advantage of producing waves with always strictly positive speed, thus avoiding resonance with the "boundary" problems at nodes v^e.

Remark 6.2.1. *We can consider a slope m, defining the flux*

$$f_\varepsilon(\rho,\mu) = \begin{cases} m\,\rho, & \rho \leq \mu, \\ m\,\mu + \varepsilon(\rho - \mu), & \rho \geq \mu, \end{cases} \tag{6.4}$$

or different slopes m^e, defining the flux

$$f_\varepsilon^e(\rho,\mu) = \begin{cases} m^e\rho, & 0 \leq \rho \leq \mu, \\ m^e\mu + \varepsilon(\rho - \mu), & \mu \leq \rho \leq \rho_{\max}^e, \end{cases} \tag{6.5}$$

where $m^e \geq 0$ represents the velocity of each processor and is given by

$$m^e = \frac{L^e}{T^e},$$

with L^e and T^e, respectively, the fixed length and processing time of processor e.

From now on, for simplicity we assume that ε is fixed and the flux is the same for each arc e; we then drop the indices and thus indicate the flux by $f(\rho,\mu)$. The general case can be treated similarly.

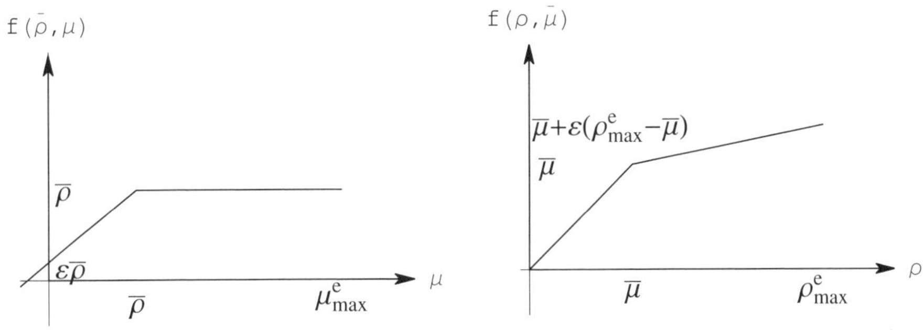

Figure 6.3. *Flux* (F)*: Left, $f(\bar\rho,\mu)$; right, $f(\rho,\bar\mu)$.*

For a scalar conservation law, a Riemann problem is a Cauchy problem for an initial datum of Heaviside type that is piecewise constant with only one discontinuity. One looks for centered solutions, i.e., $\rho(t,x) = \phi(\frac{x}{t})$ formed by simple waves, which are the building blocks to construct solutions to the Cauchy problem via the wave-front tracking algorithm. These solutions are formed by continuous waves called rarefactions and by travelling discontinuities called shocks. The speed of waves is related to the values of f'; see [10].

Analogously, we call the Riemann problem for a junction the Cauchy problem corresponding to an initial datum which is constant on each supply line.

Definition 6.1. *A Riemann Solver for the node v^e consists of a map $RS : [0,\rho^e_{\max}] \times [0,\mu^e_{\max}] \times [0,\rho^{e+1}_{\max}] \times [0,\mu^{e+1}_{\max}] \mapsto [0,\rho^e_{\max}] \times [0,\mu^e_{\max}] \times [0,\rho^{e+1}_{\max}] \times [0,\mu^{e+1}_{\max}]$ that associates with a Riemann datum $(\rho^{e,0},\mu^{e,0},\rho^{e+1,0},\mu^{e+1,0})$ at v^e a vector $(\hat{\rho}^e,\hat{\mu}^e,\hat{\rho}^{e+1},\hat{\mu}^{e+1})$ so that the solution is given by the waves $(\rho^{e,0},\hat{\rho}^e)$ and $(\mu^{e,0},\hat{\mu}^e)$ on the arc e and by the waves $(\hat{\rho}^{e+1},\rho^{e+1,0})$ and $(\hat{\mu}^{e+1},\mu^{e+1,0})$ on the arc $e+1$. We require the consistency condition*

(CC) $RS(RS(\rho^{e,0},\mu^{e,0},\rho^{e+1,0},\mu^{e+1,0})) = RS((\rho^{e,0},\mu^{e,0},\rho^{e+1,0},\mu^{e+1,0}))$.

Once a Riemann Solver is assigned we can define admissible solutions at v^e.

Definition 6.2. *Assume a Riemann Solver RS is assigned for the node v^e. Let $U = (U^e,U^{e+1})$ be such that $U^e(t,\cdot)$ and $U^{e+1}(t,\cdot)$ are of BV for every $t \geq 0$. Then U is an admissible weak solution of* (6.3) *related to RS at the junction v^e if and only if the following property holds for almost every t. Setting*

$$\widetilde{U}^e(t) = (U^e(\cdot,b^e-),U^{e+1}(\cdot,a^e+))$$

we have $RS(\widetilde{U}^e(t)) = \widetilde{U}^e(t)$.

Our aim is to solve the Cauchy problem on $[0,+\infty[$ for given initial data.

6.3 Dynamics on Arcs

Let us fix an arc e and analyze system (6.3): it is a system of conservation laws in the variables $U = (\rho,\mu)$:

$$U_t + F(U)_x = 0, \tag{6.6}$$

with flux function given by $F(U) = (f(\rho,\mu),-\mu)$; thus the Jacobian matrix of the flux is

$$DF(\rho,\mu) = \begin{cases} \begin{pmatrix} 1 & 0 \\ 0 & -1 \end{pmatrix}, & \text{if } \rho < \mu, \\[2ex] \begin{pmatrix} \varepsilon & 1-\varepsilon \\ 0 & -1 \end{pmatrix}, & \text{if } \rho > \mu. \end{cases}$$

The eigenvalues and eigenvectors are given by

$$\lambda_1(\rho,\mu) \equiv -1, \qquad r_1(\rho,\mu) = \begin{cases} \begin{pmatrix} 0 \\ 1 \end{pmatrix}, & \text{if } \rho < \mu, \\[2ex] \begin{pmatrix} -\frac{1-\varepsilon}{1+\varepsilon} \\ 1 \end{pmatrix}, & \text{if } \rho > \mu, \end{cases}$$

$$\lambda_2(\rho,\mu) = \begin{cases} 1, & \text{if } \rho < \mu, \\ \varepsilon, & \text{if } \rho > \mu, \end{cases} \qquad r_2(\rho,\mu) \equiv \begin{pmatrix} 1 \\ 0 \end{pmatrix}.$$

Hence the Hugoniot curves for the first family are vertical lines above the secant $\rho = \mu$ and lines with slope close to $-1/2$ below the same secant. The Hugoniot curves for the second family are just horizontal lines. Since we consider positive and bounded values for the variables, we fix the invariant region:

$$\mathcal{D} = \{(\rho,\mu) : 0 \le \rho \le \rho_{\max},\ 0 \le \mu \le \mu_{\max},$$
$$0 \le (1+\varepsilon)\rho + (1-\varepsilon)\mu \le (1+\varepsilon)\rho_{\max} = 2(1-\varepsilon)\mu_{\max}\};$$

see Figure 6.4.

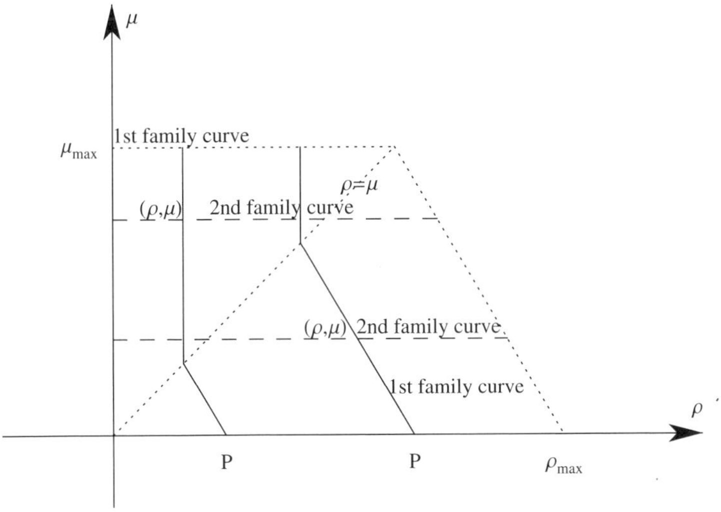

Figure 6.4. *First and second family curves.*

Observe that

$$\rho_{\max} = \mu_{\max}\frac{2}{1+\varepsilon}. \tag{6.7}$$

For future use, we compute the minimal flux along curves of the first family. That is, given an initial datum (ρ_0,μ_0), we want to determine the minimal value of the flux at points of the curve of the first family passing through (ρ_0,μ_0). From Figure 6.4, the point P, where the minimum flux is obtained, is that corresponding to $\mu = 0$. We have to distinguish two cases. Assume first $\rho_0 \le \mu_0$. Then, P is given by

$$P = (\rho_P,\mu_P) = \begin{pmatrix} \rho_0 \\ \rho_0 \end{pmatrix} + t\begin{pmatrix} 1 \\ -\frac{1+\varepsilon}{1-\varepsilon} \end{pmatrix},$$

where t is such that $\mu_P = 0$. We thus deduce

$$t = \frac{1-\varepsilon}{1+\varepsilon}\rho_0,$$

and hence

$$\rho_P = \frac{2}{1+\varepsilon}\rho_0.$$

Since $\rho_P \geq \mu_P$, the minimal flux is given by

$$f(P) = \frac{2\varepsilon}{1+\varepsilon}\rho_0.$$

Assume now $\rho_0 > \mu_0$; then P is given by

$$P = (\rho_P, \mu_P) = \begin{pmatrix} \rho_0 \\ \mu_0 \end{pmatrix} + t \begin{pmatrix} 1 \\ -\frac{1+\varepsilon}{1-\varepsilon} \end{pmatrix},$$

where t is such that $\mu_P = 0$. We thus deduce

$$t = \frac{1-\varepsilon}{1+\varepsilon}\mu_0,$$

and hence

$$\rho_P = \rho_0 + \frac{1-\varepsilon}{1+\varepsilon}\mu_0.$$

Since $\rho_P \geq \mu_P$, the minimal flux is given by

$$f(P) = \varepsilon\rho_0 + \frac{\varepsilon(1-\varepsilon)}{1+\varepsilon}\mu_0.$$

Therefore we get the following proposition.

Proposition 6.3. *Given (ρ_0, μ_0), the minimal value of the flux at points of the curve of the first family passing through (ρ_0, μ_0) is given by*

$$f_{min}((\rho_0, \mu_0)) = \begin{cases} \frac{2\varepsilon}{1+\varepsilon}\rho_0, & \text{if } \rho_0 \leq \mu_0, \\ \varepsilon\rho_0 + \frac{\varepsilon(1-\varepsilon)}{1+\varepsilon}\mu_0, & \text{if } \rho_0 > \mu_0. \end{cases}$$

Lemma 6.4. *Given an initial datum (ρ_0, μ_0), the maximum value of the density of the curve of the second family passing through (ρ_0, μ_0) and belonging to the invariant region is given by*

$$\rho_M(\mu_0) = \rho_{\max} - \mu_0 \frac{\rho_{\max} - \mu_{\max}}{\mu_{\max}}. \tag{6.8}$$

Proof. From Figure 6.4, the maximum value is obtained by the intersection of the curve of the second family passing through (ρ_0, μ_0) and the line connecting the points $(\rho_{\max}, 0)$ and $(\mu_{\max}, \mu_{\max})$:

$$\rho_M(\mu_0) = \rho_{\max} - \mu_0 \frac{\rho_{\max} - \mu_{\max}}{\mu_{\max}}.$$

From (6.7) we get

$$\rho_M(\mu_0) = \frac{2}{1+\varepsilon}\mu_{\max} - \frac{1-\varepsilon}{1+\varepsilon}\mu_0. \quad \square \tag{6.9}$$

Both families are linearly degenerate, and, in Riemann invariant coordinates, the total variation decreases along entropic solutions. Define a change of variables by

$$(\tilde{\rho},\tilde{\mu}) = \begin{pmatrix} \rho + \frac{1-\varepsilon}{1+\varepsilon}[\mu-\rho]_- \\ \mu \end{pmatrix}, \tag{6.10}$$

where $[\cdot]_- = -\min\{0,\cdot\}$. Then, in the new coordinates, the total variation decreases along wave-front tracking solutions. Notice that such a change of variables is piecewise differentiable with Jacobian matrix

$$J = \begin{cases} \begin{pmatrix} 1 & 0 \\ 0 & 1 \end{pmatrix}, & \text{if } \rho < \mu, \\ \begin{pmatrix} \frac{2\varepsilon}{1+\varepsilon} & -\frac{1-\varepsilon}{1+\varepsilon} \\ 0 & 1 \end{pmatrix}, & \text{if } \rho > \mu. \end{cases} \tag{6.11}$$

In particular, indicating by $T.V.$ the total variation in the original coordinates and $\widetilde{T.V.}$ the total variation in the new coordinates, along a wave-front tracking solution we get

$$T.V.(U(t)) \le \frac{2\varepsilon}{1+\varepsilon}\widetilde{T.V.}(U(t)) \le \frac{2\varepsilon}{1+\varepsilon}\widetilde{T.V.}(U(0)) \le \frac{2\varepsilon}{1+\varepsilon}T.V.(U(0)). \tag{6.12}$$

However, our aim is to estimate the total variation of the flux, since this is the quantity which is naturally conserved at nodes. The direct estimate on $T.V.(U)$ would allow us to estimate the flux variation, but only by $T.V.(F(U)) \le (1/\varepsilon)T.V.(U)$, which is highly not optimal.

6.4 Riemann Solvers at Nodes

In this section we discuss possible definitions of a general Riemann Solver, which conserves the flux at nodes. We fix a node v^e and a Riemann initial datum constantly equal to $(\rho^{e,0},\mu^{e,0})$ on e and constantly equal to $(\rho^{e+1,0},\mu^{e+1,0})$ on $e+1$.

First observe that the following lemmas hold.

Lemma 6.5. *On the incoming arc, only waves of the first family may be produced, while on the outgoing arc only waves of the second family may be produced.*

Lemma 6.6. *The Riemann problem at node v^e admits a solution if the following holds: if $\rho^{e,0} \le \mu^{e,0}$, then*

$$\mu^{e+1,0}(1-\varepsilon)+\varepsilon\left(\rho^{e+1}_{\max} - \frac{2}{1+\varepsilon}\rho^{e,0}\right) \ge 0. \tag{6.13}$$

If $\rho^{e,0} > \mu^{e,0}$, then

$$(1-\varepsilon)\left(\mu^{e+1,0} - \frac{\varepsilon}{1+\varepsilon}\mu^{e,0}\right) + \varepsilon(\rho^{e+1}_{\max} - \rho^{e,0}) \ge 0. \tag{6.14}$$

Proof. In order to have a solution, the minimum value of incoming flux for the node v^e must be less than the maximum value of the outgoing flux for the node v^e. Since only waves of the second family can arise on the outgoing arc, the maximum value of the outgoing flux is given by

$$\mu^{e+1,0}+\varepsilon(\rho_{\max}^{e+1}-\mu^{e+1,0}),$$

while the minimum value of the incoming flux is given by Proposition 6.3. Comparing the two, we obtain conditions (6.13) and (6.14). □

Remark 6.4.1. *Conditions* (6.13) *and* (6.14) *are fulfilled if* $\rho_{\max}^{e+1}\geq 2\rho^{e,0}$ *and* $\mu^{e+1,0}\geq\mu^{e,0}$, *which is a condition on the initial datum.*

We are now ready to describe a general solution to a Riemann problem at v^e. From Lemma 6.5, given the initial datum $(\rho^{e,0},\mu^{e,0},\rho^{e+1,0},\mu^{e+1,0})$, for every Riemann Solver it follows that

$$\begin{aligned}\hat\rho^{e}&=\varphi(\hat\mu^{e}),\\ \hat\mu^{e+1}&=\mu^{e+1,0},\end{aligned}$$

where the function $\varphi(\cdot)$ describes the first family curve through $(\rho^{e,0},\mu^{e,0})$ as a function of $\hat\mu^e$. The expression of such a curve changes at a particular value $\bar\mu^e$, given by

$$\bar\mu^{e}=\begin{cases}\rho^{e,0}, & \text{if } \rho^{e,0}\leq\mu^{e,0},\\ \frac{1+\varepsilon}{2}\rho^{e,0}+\frac{1-\varepsilon}{2}\mu^{e,0}, & \text{if } \rho^{e,0}>\mu^{e,0}.\end{cases}\tag{6.15}$$

Let us now discuss how $\hat\rho^{e+1}$ and $\hat\mu^e$ can be chosen. The conservation of flux at the node can be written as

$$f(\varphi(\hat\mu^{e}),\hat\mu^{e})=f(\hat\rho^{e+1},\mu^{e+1,0}).\tag{6.16}$$

We have to distinguish two cases.

Case α $\mu^{e+1,0}<\bar\mu^e$;

Case β $\bar\mu^e\leq\mu^{e+1,0}$.

In both cases $\bar\mu^e$ and $\mu^{e+1,0}$ individuate in the plane $(\hat\rho^{e+1},\hat\mu^e)$ four regions, A, B, C, D, so defined:

$$\begin{aligned}A&=\{(\hat\rho^{e+1},\hat\mu^{e}):0\leq\hat\rho^{e+1}\leq\mu^{e+1,0},\bar\mu^{e}\leq\hat\mu^{e}\leq\mu^{e}_{\max}\};\\ B&=\{(\hat\rho^{e+1},\hat\mu^{e}):\mu^{e+1,0}\leq\hat\rho^{e+1}\leq\rho^{e+1}_{\max},\bar\mu^{e}\leq\hat\mu^{e}\leq\mu^{e}_{\max}\};\\ C&=\{(\hat\rho^{e+1},\hat\mu^{e}):0\leq\hat\rho^{e+1}\leq\mu^{e+1,0},0\leq\hat\mu^{e}\leq\bar\mu^{e}\};\\ D&=\{(\hat\rho^{e+1},\hat\mu^{e}):\mu^{e+1,0}\leq\hat\rho^{e+1}\leq\rho^{e+1}_{\max},0\leq\hat\mu^{e}\leq\bar\mu^{e}\}.\end{aligned}$$

Case 1 $(\hat\rho^{e+1},\hat\rho^e)$ belongs to region A.

The relation (6.16) becomes

$$\hat\rho^{e+1}=\bar\mu^{e}.\tag{6.17}$$

In Case α (6.17) cannot hold, since $\mu^{e+1,0}<\bar\mu^e$ and in region A we have $0\leq\hat\rho^{e+1}\leq\mu^{e+1,0}$, drawn in Figure 6.5.

In Case β, since $\bar\mu^e\leq\mu^{e+1,0}$, (6.17) is satisfied along the line $\hat\rho^{e+1}=\bar\mu^e$, drawn in Figure 6.6.

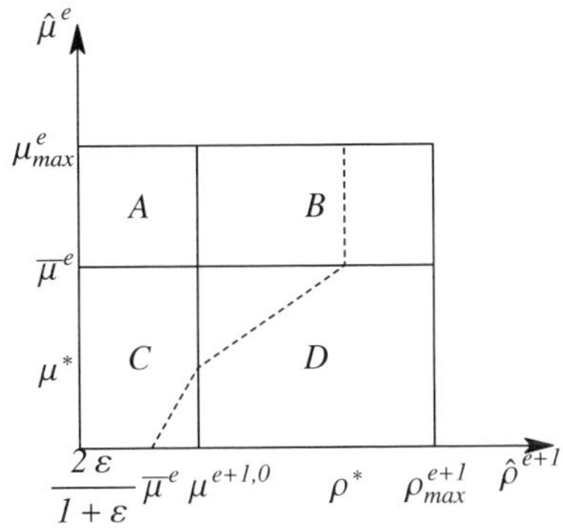

Figure 6.5. *Case* $\alpha : \mu^{e+1,0} < \bar{\mu}^e$.

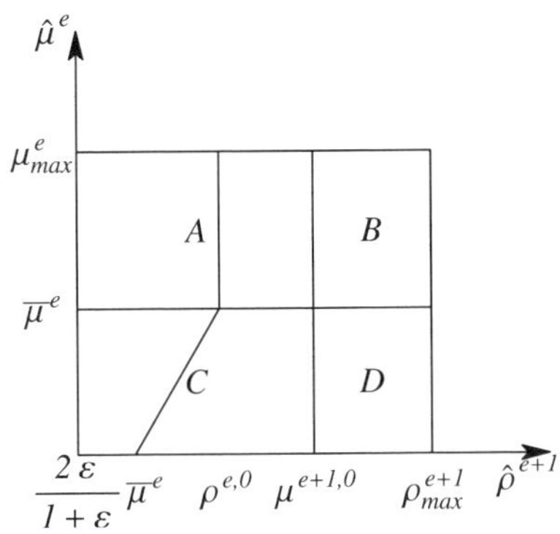

Figure 6.6. *Case* $\beta : \bar{\mu}^e \leq \mu^{e+1,0}$.

Case 2 $(\hat{\rho}^{e+1}, \hat{\mu}^e)$ belongs to region B.

The relation (6.16) reads

$$\mu^{e+1,0} + \varepsilon(\hat{\rho}^{e+1} - \mu^{e+1,0}) = \bar{\mu}^e, \tag{6.18}$$

from which we obtain that

$$\hat{\rho}^{e+1} = \frac{\bar{\mu}^e - (1-\varepsilon)\mu^{e+1,0}}{\varepsilon} = \rho^*.$$

Equation (6.18) admits solutions only if

$$\mu^{e+1,0} \leq \rho^* \leq \rho^{e+1}_{\max}.$$

The first inequality is equivalent to $\mu^{e+1,0} \leq \bar{\mu}^e$, which, in Case α, always holds and, in Case β, holds only if equality holds and we find back solutions of region A.

The relation $\rho^* \leq \rho_{\max}^{e+1}$ is satisfied if and only if

$$\bar{\mu}^e \leq (1-\varepsilon)\mu^{e+1,0} + \varepsilon\rho_{\max}^{e+1},$$

and the latter holds if $\varepsilon\rho_{\max}^{e+1}$ is big enough. In that case (6.18) admits solutions given by $\hat{\rho}^{e+1} = \rho^*$ (see dashed line in Figure 6.5).

Case 3 $(\hat{\rho}^{e+1}, \hat{\mu}^e)$ belongs to region C.

To satisfy (6.16), it must hold that

$$\hat{\rho}^{e+1} = (1-\varepsilon)\hat{\mu}^e + \varepsilon\hat{\rho}^e. \tag{6.19}$$

Since $\hat{\mu}^e \leq \bar{\mu}^e$,

$$\hat{\rho}^e = \varphi(\hat{\mu}^e) = \bar{\mu}^e - \frac{1-\varepsilon}{1+\varepsilon}(\hat{\mu}^e - \bar{\mu}^e).$$

After straightforward computations, (6.19) is equivalent to

$$\hat{\rho}^{e+1} = \frac{1-\varepsilon}{1+\varepsilon}\hat{\mu}^e + \frac{2}{1+\varepsilon}\bar{\mu}^e.$$

In Case β such an equality is satisfied along the line connecting the points $(\frac{2\varepsilon}{1+\varepsilon}\bar{\mu}^e, 0)$ and $(\bar{\mu}^e, \bar{\mu}^e)$; see Figure 6.6.

In Case α solutions exist only under some conditions. In fact, for $\hat{\mu}^e = 0$ or for $\hat{\rho}^{e+1} = \mu^{e+1,0}$, the point is admissible only if

$$\frac{2\varepsilon}{1+\varepsilon}\bar{\mu}^e \leq \mu^{e+1,0}.$$

Case 4 $(\hat{\rho}^{e+1}, \hat{\mu}^e)$ belongs to region D.

The relation (6.16) reads

$$(1-\varepsilon)\mu^{e+1,0} + \varepsilon\hat{\rho}^{e+1} = (1-\varepsilon)\hat{\mu}^e + \varepsilon\hat{\rho}^e. \tag{6.20}$$

The expression of φ is as above. Then, after straightforward computations, we get the equivalent equation:

$$\hat{\rho}^{e+1} = \frac{1}{\varepsilon}\left(\frac{1-\varepsilon}{1+\varepsilon}\hat{\mu}^e + \frac{2\varepsilon}{1+\varepsilon}\bar{\mu}^e - (1-\varepsilon)\mu^{e+1,0}\right). \tag{6.21}$$

If $\hat{\mu}^e = 0$, then (6.21) becomes

$$\hat{\rho}^{e+1} = \frac{1}{\varepsilon}\left(\frac{2\varepsilon}{1+\varepsilon}\bar{\mu}^e - (1-\varepsilon)\mu^{e+1,0}\right).$$

The condition $\hat{\rho}^{e+1} \geq \mu^{e+1,0}$ is always false in Case β.

If $\hat{\mu}^e = \bar{\mu}^e$, then (6.21) becomes

$$\hat{\rho}^{e+1} = \frac{1}{\varepsilon}\left(\bar{\mu}^e - (1-\varepsilon)\mu^{e+1,0}\right). \tag{6.22}$$

In Case β the point is admissible only if $\bar{\mu}^e = \mu^{e+1,0}$ and we find back a point of region A.

In case α there are solutions only under some conditions. Long but standard computations permit us to find back conditions of Lemma 6.6.

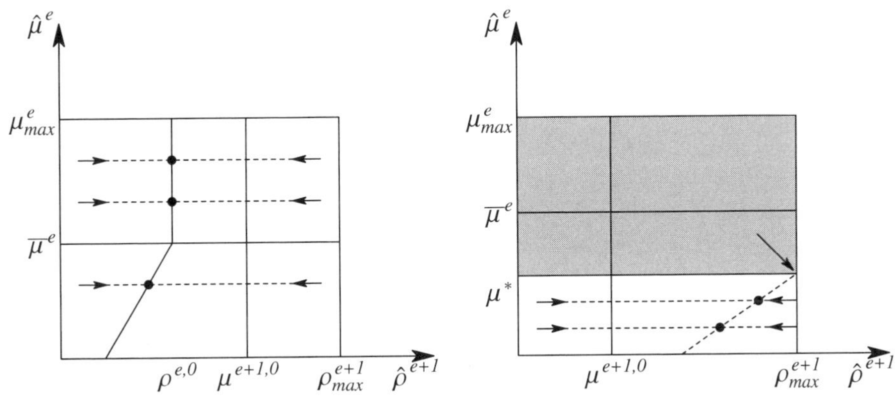

Figure 6.7. *An example of a Riemann Solver: Case β (on the left) and Case α (on the right).*

6.5 A Riemann Solver According to Rule SC1

A geometrically natural Riemann Solver is the following. In Case β we can define a Riemann Solver mapping every initial datum on the line $\hat{\mu}^e = c$ to the intersection of the same line with that drawn in Figure 6.6.

In Case α it may happen that there is no admissible solution on a given line $\hat{\mu}^e = c$. Therefore, we can use the same procedure if the line $\hat{\mu}^e = c$ intersects the dashed line of Figure 6.5 while mapping all other points to the admissible solution with the highest value of $\hat{\mu}^e$.

The obtained Riemann Solver is depicted in Figure 6.7 and satisfies the policy SC1. On the left, there is Case β with all points mapped horizontally, while, on the right, there is Case α: all points of the white region are mapped horizontally and all points of the dark region are mapped to the point indicated by the arrow.

Remark 6.5.1. *If $\hat{\rho}^{e+1} \leq \mu^{e+1,0}$, then the solution $(\hat{\rho}^{e+1}, \rho^{e+1,0})$ is a contact discontinuity. The same happens if $\hat{\rho}^{e+1} \geq \mu^{e+1,0}$ and $\hat{\rho}^{e+1} > \mu^{e+1,0}$. If $\hat{\rho}^{e+1} > \mu^{e+1,0}$ and $\rho^{e+1,0} < \mu^{e+1,0}$, the solution consists of two contact discontinuities.*

Remark 6.5.2. *If we choose the flux*

$$f(\rho,\mu) = \begin{cases} \rho, & 0 \leq \rho \leq \mu, \\ \mu - \varepsilon(\rho - \mu), & \mu \leq \rho \leq \rho^e_{\max}, \end{cases}$$

we have no solution in some cases.

Let us define in detail the Riemann Solver described in Figure 6.7. We introduce the notations

$$\gamma^e_{\max} = f(\rho^e_{\max}, \mu^{e,0}),$$
$$\gamma^{e+1}_{\max} = f(\rho^{e+1}_{\max}, \mu^{e+1,0}).$$

Proposition 6.7. *Fix a node v^e. There exists a unique vector $(\hat{\rho}^e, \hat{\mu}^e, \hat{\rho}^{e+1}, \hat{\mu}^{e+1})$ for every Riemann initial datum $(\rho^{e,0}, \mu^{e,0}, \rho^{e+1,0}, \mu^{e+1,0})$ at v^e such that the following hold:*

(a) *if $f(\rho^{e,0}, \mu^{e,0}) \leq \gamma_{\max}^{e+1}$, then*

$$\hat{\mu}^e = \mu^{e,0}, \quad \hat{\mu}^{e+1} = \mu^{e+1,0},$$

$$\hat{\rho}^e = \rho^{e,0},$$

$$\hat{\rho}^{e+1} = \begin{cases} f(\rho^{e,0}, \mu^{e,0}), & \text{if } f(\rho^{e,0}, \mu^{e,0}) \leq \mu^{e+1,0}, \\ \frac{f(\rho^{e,0}, \mu^{e,0}) - \mu^{e+1,0}}{\varepsilon} + \mu^{e+1,0}, & \text{if } \mu^{e+1,0} \leq f(\rho^{e,0}, \mu^{e,0}) \leq \gamma_{\max}^{e+1}; \end{cases}$$

(b) *if $f(\rho^{e,0}, \mu^{e,0}) > \gamma_{\max}^{e+1}$, then*

$$\hat{\rho}^e = \rho^{e,0}, \quad \hat{\rho}^{e+1} = \rho_{\max}^{e+1},$$

$$\hat{\mu}^e = \frac{\gamma_{\max}^{e+1} - \varepsilon \rho^{e,0}}{1 - \varepsilon}, \quad \hat{\mu}^{e+1} = \mu^{e+1,0}.$$

Proof. The proof of case (a) is obvious. Let us consider case (b). Starting from the values $\hat{\rho}^e, \hat{\rho}^{e+1}, \hat{\mu}^{e+1}$ defined in the statement, let us determine $\hat{\mu}^e$. We distinguish two cases:

(b1) $\mu^{e,0} \leq \rho^{e,0}$ ($\rho^{e,0}$ belongs to the region with slope ε of $f(\rho, \mu^{e,0})$);

(b2) $\mu^{e,0} > \rho^{e,0}$ ($\rho^{e,0}$ belongs to the region with slope 1 of $f(\rho, \mu^{e,0})$).

In case (b1), $\mu^{e,0}$ belongs to the region of slope $1 - \varepsilon$ of $f(\rho^{e,0}, \cdot)$. Since $\gamma_{\max}^{e+1} < f(\rho^{e,0}, \mu^{e,0})$, it follows that there exists a unique $\hat{\mu}^e$ such that $f(\rho^{e,0}, \hat{\mu}^e) = \gamma_{\max}^{e+1}$. Moreover, $\hat{\mu}^e < \mu^{e,0} \leq \rho^{e,0}$, from which we get that $\rho^{e,0}$ belongs to the region with slope ε of $f(\cdot, \hat{\mu}^e)$.

Let us analyze case (b2). Since $\gamma_{\max}^{e+1} < f(\rho^{e,0}, \mu^{e,0})$, it follows that there exists a unique $\hat{\mu}^e$ such that $f(\rho^{e,0}, \hat{\mu}^e) = \gamma_{\max}^{e+1}$. Moreover, $\hat{\mu}^e < \mu^{e,0}$. In fact, $\hat{\mu}^e > \mu^{e,0}$ leads to $\gamma_{\max}^{e+1} \geq f(\rho^{e,0}, \mu^{e,0})$. It follows that $\rho^{e,0}$ belongs to the region with slope ε of $f(\cdot, \hat{\mu}^e)$. ☐

Theorem 6.8. *The Riemann Solver described in Proposition 6.7 is in accordance with rule SC1.*

Proof. In case (a), since $f(\rho^{e,0}, \mu^{e,0}) \leq \gamma_{\max}^{e+1}$, rule SC1 implies that only a wave of the second family in the outgoing arc $e+1$ is created.

In case (b), $f(\rho^{e,0}, \mu^{e,0}) > \gamma_{\max}^{e+1}$ and rule SC1 implies that the maximal wave of the second family in the outgoing arc $e+1$ is created. Therefore $\hat{\rho}^{e+1} = \rho_{\max}^{e+1}$. The value of $\hat{\mu}^e$ then follows from the conservation of the flux. ☐

Let us pass now to consider the solvability of Riemann problems according to the Riemann Solver above.

Lemma 6.9. *Consider a supply chain on which the initial datum verifies $\mu^{e,0} = \mu_{\max}^e$; i.e., the production rate is at its maximum. A sufficient condition for the solvability of all Riemann problems, according to rule SC1, on the supply chain at every time is*

$$\rho_{\max}^{e+2} \geq \rho_{\max}^e \quad \forall e.$$

Proof. Note that the lower the value $\mu^{e+1,0}$, the more difficult it is to find a solution. Thus, we must consider the lowest value $\tilde{\mu}$ produced by the node v^{e+1}, which is given by

$$\tilde{\mu} + \varepsilon(\rho_{\max}^{e+1} - \tilde{\mu}) = \varepsilon \rho_{\max}^{e+2}.$$

It follows that

$$\tilde{\mu}(1-\varepsilon) = \varepsilon(\rho_{\max}^{e+2} - \rho_{\max}^{e+1}).$$

Now the worst case at node v^e happens if $\rho^{e,0} = \rho_{\max}^{e}$ and $\mu^{e+1,0} = \tilde{\mu}$. The solvability condition, in this case, reads as

$$\varepsilon \rho_{\max}^{e} \leq \tilde{\mu} + \varepsilon(\rho_{\max}^{e+1} - \tilde{\mu}) = \varepsilon \rho_{\max}^{e+2};$$

then we conclude. □

6.6 Estimates on Density Variation for the Riemann Solver SC1

In this section, we derive estimates on the total variation of the flux, densities, and service frequencies along a wave-front tracking approximate solution (constructed as in [17]). This allows us to construct the solutions to the Cauchy problem in the standard way; see [10]. From now on, we fix a supply chain and a wave-front tracking approximate solution (ρ, μ), defined on the supply chain. We assume that the initial datum verifies $T.V.(\mu(0)) = 0$, so that no wave of the first family is present at initial time.

Flux Estimates inside Arcs

In this section, we show how to estimate the flux variation for interaction of waves inside arcs.

Proposition 6.10. *Assume that a second family wave $((\rho_l, \mu_l), (\rho_m, \mu_m))$ interacts with a first family wave $((\rho_m, \mu_m), (\rho_r, \mu_r))$. If $\mu_r < \mu_m$, then the flux variation decreases.*

Proof. By assumption, $\mu_m = \mu_l$.

Let us consider first the case in which $\rho_m < \rho_l$. Assume first that both left and right data are on the secant $\rho = \mu$; thus $\rho_l = \mu_l$ and $\rho_r = \mu_r$. In this case we also have $\rho_m = \rho_r$; see Figure 6.8(left). Then, before the interaction the flux variation is given by

$$T.V.(f)^{-} = \rho_l - \rho_r.$$

In fact, the flux does not change along the first family wave. After the interaction, a first family wave connecting (ρ_l, μ_l) to (ρ^*, μ_r) and a second family wave connecting (ρ^*, μ_r) to (ρ_r, μ_r) are produced, where

$$\rho^* = \rho_l + \frac{1-\varepsilon}{1+\varepsilon}(\mu_l - \mu_r) = \rho_l + \frac{1-\varepsilon}{1+\varepsilon}(\rho_l - \rho_r); \tag{6.23}$$

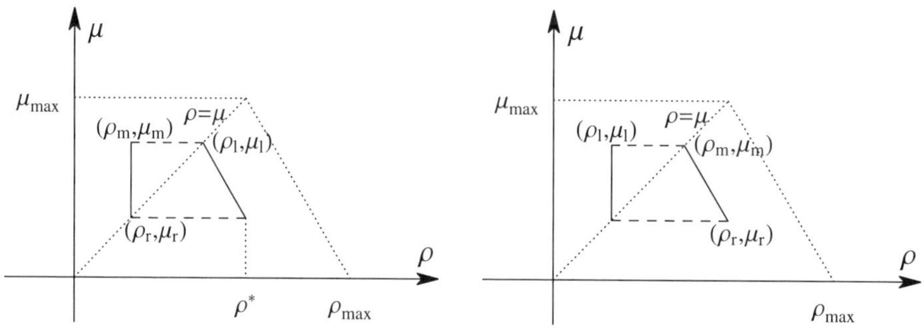

Figure 6.8. *Interactions inside an arc.*

see Figure 6.8(left). Thus, the flux total variation after the interaction is given by

$$\begin{aligned} T.V.(f)^+ &= \left|\mu_l+\varepsilon(\rho_l-\mu_l)-\big(\mu_r+\varepsilon(\rho^*-\mu_r)\big)\right| \\ &\quad + \left|\mu_r+\varepsilon(\rho^*-\mu_r)-\big(\mu_r+\varepsilon(\rho_r-\mu_r)\big)\right| \\ &= \left|\mu_l-\mu_r-\varepsilon\big(\rho^*-\mu_r\big)\right|+\varepsilon\left|\rho^*-\mu_r\right| = I_1+I_2. \end{aligned}$$

Using (6.23), $\rho_l=\mu_l$, and $\rho_r=\mu_r$, we get

$$I_1 = \left|\rho_l-\rho_r-\varepsilon\left(\rho_l+\frac{1-\varepsilon}{1+\varepsilon}(\rho_l-\rho_r)-\rho_r\right)\right|$$

$$= \left|(\rho_l-\rho_r)\left(1-\varepsilon\left(1+\frac{1-\varepsilon}{1+\varepsilon}\right)\right)\right| = (\rho_l-\rho_r)\frac{1-\varepsilon}{1+\varepsilon},$$

while

$$I_2 = \varepsilon\left|\rho_l+\frac{1-\varepsilon}{1+\varepsilon}(\rho_l-\rho_r)-\rho_r\right|$$

$$= \varepsilon(\rho_l-\rho_r)\left(1+\frac{1-\varepsilon}{1+\varepsilon}\right) = (\rho_l-\rho_r)\frac{2\varepsilon}{1+\varepsilon}.$$

Finally, summing up, it follows that

$$T.V.(f)^+ = I_1+I_2 = \rho_l-\rho_r = T.V.(f)^-.$$

Let us now pass to the general case; see Figure 6.9. We can consider the points $\tilde{U}_l=(\mu_l,\mu_l)$, $\tilde{U}_r=(\rho_m,\rho_m)$, and $U^*=\tilde{U}_l+(\mu_l-\rho_m)(\frac{1-\varepsilon}{1+\varepsilon},1)$; see Figure 6.9. Then we can split the estimate, first noticing that the flux variation along waves $(\tilde{U}_l,(\rho_m,\mu_m))$ and $((\rho_m,\mu_m),\tilde{U}_r)$ is equivalent to the flux along waves $(\tilde{U}_l,U^*)$ and $(U^*,\tilde{U}_r)$. It remains to estimate the flux variation along waves all in the region where $\rho>\mu$. But in this region the flux variation does not change if we translate waves along the first or the second eigenvector; thus we are finished.

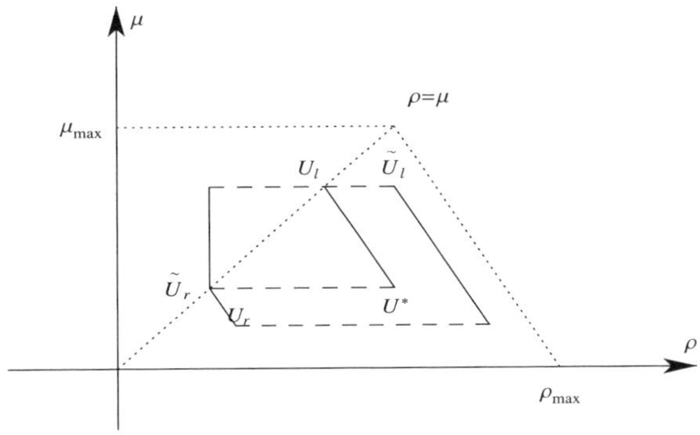

Figure 6.9. *Interactions inside an arc.*

Assume now $\rho_m > \rho_l$. First, we suppose $\rho_m = \mu_m$ and $\mu_r = \rho_l$; see Figure 6.8(right). Then before the interaction the flux variation is given by

$$T.V.(f)^- = |\rho_m - \rho_l| + |\mu_r + \varepsilon(\rho_r - \mu_r) - \rho_m|$$

$$= |\rho_m - \rho_l| + |(1-\varepsilon)\rho_l + \varepsilon\rho_r - \rho_m| = I_1 + I_2.$$

Moreover, we can compute

$$\rho_r = \rho_m + \frac{1-\varepsilon}{1+\varepsilon}(\mu_l - \mu_r) = \rho_m + \frac{1-\varepsilon}{1+\varepsilon}(\rho_m - \rho_l); \tag{6.24}$$

see Figure 6.8(right). Then we get

$$I_2 = \left|(1-\varepsilon)\rho_l + \varepsilon\left(\rho_m + \frac{1-\varepsilon}{1+\varepsilon}(\rho_m - \rho_l) - \rho_m\right)\right|$$

$$= \left|\rho_m\left(1 - \varepsilon - \varepsilon\frac{1-\varepsilon}{1+\varepsilon}\right) + \rho_l\left(-1 + \varepsilon + \varepsilon\frac{1-\varepsilon}{1+\varepsilon}\right)\right|$$

$$= (\rho_m - \rho_l)\frac{1-\varepsilon}{1+\varepsilon}.$$

Hence

$$T.V.(f)^- = I_1 + I_2 = (\rho_m - \rho_l)\frac{2}{1+\varepsilon}.$$

After the interaction, a first family wave connecting (ρ_l, μ_l) to (ρ_l, μ_r) and a second family wave connecting (ρ_l, μ_r) to (ρ_r, μ_r) are produced. Thus the flux total variation after the interaction is given by

$$T.V.(f)^+ = |\mu_r + \varepsilon(\rho_r - \mu_r) - \rho_l|.$$

In fact, the flux does not change along the first family wave. Using $\mu_r = \rho_l$, we get

$$T.V.(f)^+ = |(1-\varepsilon)\rho_l + \varepsilon\rho_r - \rho_l| = \varepsilon|\rho_r - \rho_l|.$$

By (6.24),

$$T.V.(f)^{+} = \left|\rho_m + \frac{1-\varepsilon}{1+\varepsilon}(\rho_m - \rho_l), -\rho_l\right|$$

$$= \left|\rho_m\left(1 + \frac{1-\varepsilon}{1+\varepsilon}\right) - \rho_l\left(-1 - \frac{1-\varepsilon}{1+\varepsilon}\right)\right| = \frac{2\varepsilon}{1+\varepsilon}(\rho_m - \rho_l) = \varepsilon T.V.(f)^{-}.$$

The general case is treated as above. □

The necessity of the assumption $\mu_r < \mu_l$ in Proposition 6.10 is shown by the next example.

Example 6.6.1. *Consider again an interaction between a second family wave* $((\rho_l, \mu_l), (\rho_m, \mu_m))$ *and a first family wave* $((\rho_m, \mu_m), (\rho_r, \mu_r))$.

We assume $\rho_l > \mu_l$, $\rho_m = \mu_m$, *and* $\mu_r > \rho_r$. *In particular, Proposition* 6.10 *does not apply. Then we also have* $\mu_m = \mu_l$ *and* $\rho_r = \rho_m$; *see Figure* 6.10. *Before the interaction the flux variation is given by*

$$T.V.(f)^{-} = \left|\mu_l + \varepsilon(\rho_l - \mu_l) - \left(\mu_m + \varepsilon(\rho^m - \mu_m)\right)\right| = \varepsilon|\rho_l - \rho_r|.$$

In fact, the flux does not change along the first family wave.

After the interaction, a first family wave connecting (ρ_l, μ_l) *to* (ρ^*, μ^*) *and a second family wave connecting* (ρ^*, μ^*) *to* (ρ_r, μ_r) *are produced, where*

$$\rho^* = \mu^*, \qquad \rho^* = \rho_l - \frac{1-\varepsilon}{1+\varepsilon}(\mu^* - \mu_l);$$

see Figure 6.10. *From these equalities it follows that*

$$\rho^* = \mu^* = \frac{1+\varepsilon}{2}\rho_l + \frac{1-\varepsilon}{2}\mu_l. \tag{6.25}$$

The flux variation after the interaction is

$$T.V.(f)^{+} = \left|\mu_l + \varepsilon(\rho_l - \mu_l) - \rho^*\right| + |\rho^* - \rho_m| = I_1 + I_2.$$

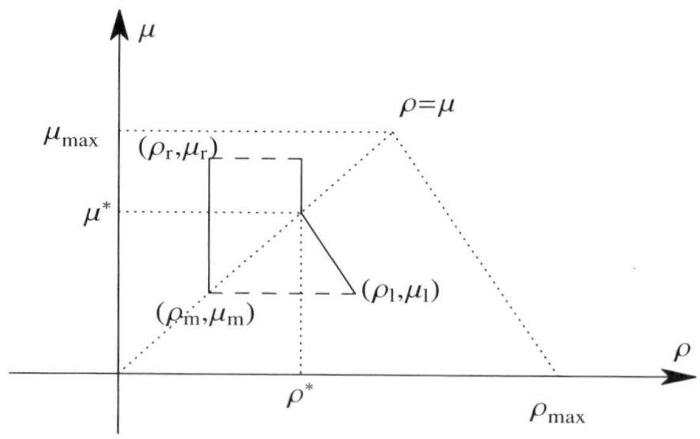

Figure 6.10. *Interactions inside an arc.*

Using (6.25) *and* $\mu_l = \rho_m$, *we obtain*

$$I_1 = \left|\rho_m + \varepsilon(\rho_l - \rho_m) - \left(\frac{1+\varepsilon}{2}\rho_l + \frac{1-\varepsilon}{2}\rho_m\right)\right|$$

$$= \left|\rho_l\left(\varepsilon - \frac{1+\varepsilon}{2}\right) + \rho_m\left(1 - \varepsilon - \frac{1-\varepsilon}{2}\right)\right| = (\rho_l - \rho_m)\frac{1-\varepsilon}{2},$$

while

$$I_2 = \left|\frac{1+\varepsilon}{2}\rho_l + \frac{1-\varepsilon}{2}\rho_m - \rho_m\right| = \frac{1+\varepsilon}{2}(\rho_l - \rho_m).$$

Finally, summing up, it follows that

$$T.V.(f)^+ = I_1 + I_2 = \rho_l - \rho_m = \frac{1}{\varepsilon}T.V.(f)^-.$$

Flux Estimates at Nodes

We now show that the flux total variation is conserved for interaction of waves with a junction. Fix a node v^e, and distinguish two cases:

(1) a wave ρ interacts with the node v^e;

(2) a wave μ interacts with the node v^e.

Assume the interaction happens at some time $\bar{t}$ and let $(\rho^{e,-}, \rho^{e+1,-})$, $(\mu^{e,-}, \mu^{e+1,-})$ and $(\rho^{e,+}, \rho^{e+1,+})$, $(\mu^{e,+}, \mu^{e+1,+})$ indicate the equilibrium configurations at the junction v^e before and after the interaction, respectively. In general $-$ and $+$ denote the values before and after the interaction, respectively. For instance $T.V.(f)^{\pm} = T.V.(f(\rho(\bar{t}\pm,\cdot),\mu(\bar{t}\pm,\cdot)))$ are the flux variations of waves before and after the interaction, respectively.

Since $(\rho^{e,-}, \mu^{e,-}, \rho^{e+1,-}, \mu^{e+1,-})$ and $(\rho^{e,+}, \mu^{e,+}, \rho^{e+1,+}, \mu^{e+1,+})$ are equilibrium configurations, the following relations hold:

$$\begin{aligned} f(\rho^{e,-}, \mu^{e,-}) &\le \mu^{e+1,-} + \varepsilon(\rho^{e+1}_{\max} - \mu^{e+1,-}),\\ f(\rho^{e,-}, \mu^{e,-}) &= f(\rho^{e+1,-}, \mu^{e+1,-}),\\ f(\rho^{e,+}, \mu^{e,+}) &\le \mu^{e+1,+} + \varepsilon(\rho^{e+1}_{\max} - \mu^{e+1,+}),\\ f(\rho^{e,+}, \mu^{e,+}) &= f(\rho^{e+1,+}, \mu^{e+1,+}). \end{aligned} \tag{6.26}$$

Let us start considering case (1) and let ρ^e be the value on the left of the wave arriving at v^e. We have two subcases:

(1.1) $f(\rho^e, \mu^{e,-}) < f(\rho^{e,-}, \mu^{e,-})$;

(1.2) $f(\rho^e, \mu^{e,-}) > f(\rho^{e,-}, \mu^{e,-})$.

In case (1.1), the new equilibrium flux $f(\rho^e, \mu^{e,-})$ is lower than $f(\rho^{e+1,-}, \mu^{e+1,-})$. Hence, only a ρ wave is produced on the outgoing arc $e+1$: this corresponds to case (a) of Proposition 6.7. We have that

$$\rho^{e,+} = \rho^e, \quad \mu^{e,+} = \mu^{e,-}, \quad \mu^{e+1,+} = \mu^{e+1,-},$$

and $\rho^{e+1,+}$ is determined by the condition

$$f(\rho^{e+1,+},\mu^{e+1,-}) = f(\rho^e,\mu^{e,-}).$$

Therefore, we get

$$T.V.(f)^- = f(\rho^{e,-},\mu^{e,-}) - f(\rho^e,\mu^{e,-})$$
$$= f(\rho^{e+1,-},\mu^{e+1,-}) - f(\rho^{e+1,+},\mu^{e+1,-}) = T.V.(f)^+.$$

In case (1.2), there are two possible situations:

(1.2.1) $f(\rho^e,\mu^{e,-}) < \mu^{e+1,-} + \varepsilon(\rho^{e+1}_{\max} - \mu^{e+1,-})$;

(1.2.2) $f(\rho^e,\mu^{e,-}) > \mu^{e+1,-} + \varepsilon(\rho^{e+1}_{\max} - \mu^{e+1,-})$.

In case (1.2.1), still only a ρ wave is produced on the outgoing arc $e+1$, and we conclude as for case (1.1).

In case (1.2.2), both a ρ wave is produced on the outgoing arc $e+1$ and a μ wave is produced on the incoming arc e. This corresponds to case (b) of Proposition 6.7, and the following relations hold:

$$\rho^{e,+} = \rho^e, \quad \mu^{e,+} < \mu^{e,-}, \quad \rho^{e+1,+} > \rho^{e+1,-}.$$

Now, from (6.26), we deduce that

$$T.V.(f)^+ = f(\rho^{e,+},\mu^{e,-}) - f(\rho^{e,+},\mu^{e,+}) + f(\rho^{e+1,+},\mu^{e+1,+}) - f(\rho^{e+1,-},\mu^{e+1,-})$$
$$= f(\rho^e,\mu^{e,-}) - f(\rho^{e,+},\mu^{e,+}) + f(\rho^{e,+},\mu^{e,+}) - f(\rho^{e,-},\mu^{e,-})$$
$$= f(\rho^e,\mu^{e,-}) - f(\rho^{e,-},\mu^{e,-}) = T.V.(f)^-.$$

Consider now case (2) and let μ^{e+1} be the value on the right of the wave. Since $T.V.(\mu(0)) = 0$ at initial time, such a wave was produced by some node; hence $\mu^{e+1} < \mu^{e+1,-}$. In particular,

$$f(\rho^{e+1,-},\mu^{e+1}) < f(\rho^{e+1,-},\mu^{e+1,-}).$$

We have two subcases:

(2.1) $f(\rho^{e,-},\mu^{e,-}) \le \mu^{e+1} + \varepsilon(\rho^{e+1}_{\max} - \mu^{e+1})$;

(2.2) $f(\rho^{e,-},\mu^{e,-}) > \mu^{e+1} + \varepsilon(\rho^{e+1}_{\max} - \mu^{e+1})$.

In case (2.1), the equilibrium flux is still $f(\rho^{e,-},\mu^{e,-})$, and only a ρ wave is produced on the outgoing arc $e+1$: this corresponds to case (a) of Proposition 6.7. The conservation of flux variation follows similarly as for case (1.1).

In case (2.2), both a ρ wave is produced on the outgoing arc $e+1$ and a μ wave is produced on the incoming arc e. This corresponds to case (b) of Proposition 6.7, and the following relations hold:

$$\rho^{e,+} = \rho^{e,-}, \quad \mu^{e,+} < \mu^{e,-}, \quad \rho^{e+1,+} > \rho^{e+1,-}.$$

Now, from (6.26) we deduce that

$$T.V.(f)^{+} = f(\rho^{e,+},\mu^{e,-}) - f(\rho^{e,+},\mu^{e,+}) + f(\rho^{e+1,+},\mu^{e+1}) - f(\rho^{e+1,-},\mu^{e+1})$$

$$= f(\rho^{e+1,-},\mu^{e+1,-}) - f(\rho^{e+1,+},\mu^{e+1}) + f(\rho^{e+1,+},\mu^{e+1}) - f(\rho^{e+1,-},\mu^{e+1})$$

$$= f(\rho^{e+1,-},\mu^{e+1,-}) - f(\rho^{e+1,-},\mu^{e+1}) = T.V.(f)^{-}.$$

Finally, we obtain the following proposition.

Proposition 6.11. *Consider a wave-front tracking solution with initial data verifying* $T.V.(\mu(0)) = 0$. *Then*

$$T.V.(f(t)) \leq T.V.(f(0+)).$$

Proof. Since $T.V.(\mu(0)) = 0$ at initial time, all μ waves are produced by some node; hence for any wave (μ_l, μ_r) we have $\mu_l > \mu_r$.

By the above analysis, $T.V.(f)$ does not increase for interactions of waves with nodes. By Proposition 6.10, $T.V.(f)$ decreases for interactions of waves inside an arc; thus we conclude. □

Estimates on Production Rate Variation

Let us indicate by $T.V.(\mu(t), e)$ the total variation of the function μ created by v^e over the interval e. The following preliminary lemma is true.

Lemma 6.12. *Assuming that* $T.V.(\mu(0)) = 0$, *a global estimate holds:*

$$T.V.(\mu(t), e) \leq \mu(0, b^e-).$$

Proof. To prove the lemma, it is enough to observe that waves produced by the Riemann Solver at nodes are always decreasing in μ. □

Observe that the estimate

$$T.V.(\mu)^{+} \leq C\, T.V.(f)^{-}$$

is false in general. Consider an interaction of type (1.2.2); thus $\mu^{e,+} < \rho^{e,-}$, and we suppose that $\mu^{e,-} > \rho^{e} > \rho^{e,-}$. Then,

$$T.V.(\mu)^{+} = |\mu^{e,-} - \mu^{e,+}| > |f(\rho^{e},\mu^{e,+}) - f(\rho^{e,-},\mu^{e,-})| = T.V.(f)^{-}.$$

If $\rho^{e,0} < \mu^{e,0}$, the estimate

$$T.V.(\mu)^{+} \leq C\, T.V.(\mu)^{-} \tag{6.27}$$

may not hold. In fact, let $\mu^{e,-} = \rho^{e}_{\max}, \rho^{e,-} = \bar{\rho}$ with $\bar{\rho}$ near to zero, and let $\mu^{e+1,0} = \rho^{e,-} = \bar{\rho}, \rho^{e+1,-} = \bar{\rho}$.

Suppose that $\mu^{e+1} = \mu^{e+1,-} - \varepsilon$. We have that $T.V.(\mu)^- = |\mu^{e+1,-} - \mu^{e+1}| = \varepsilon$, $T.V.(\mu)^+ = |\mu^{e,-} - \mu^{e+1}| = \rho^e_{\max} - \bar{\rho} - \varepsilon$. We can choose $\bar{\rho}$ so that $\rho^e_{\max} - \bar{\rho} > 2\varepsilon$; thus we are finished.

Even if an estimate on the μ variation of type (6.27) does not hold in general, we can prove the following lemma.

Lemma 6.13. *If $\rho^{e,-} \geq \mu^{e,-}$, then $T.V.(\mu)^+ \leq T.V.(\mu)^-$.*

Proof. Since we are in an equilibrium configuration, we have that

$$f(\rho^{e,-}, \mu^{e,-}) = f(\rho^{e+1,-}, \mu^{e+1,-}).$$

Observe that $\mu^{e+1} < \mu^{e+1,-}$. If a μ wave is produced, we are in the case (b) of Proposition 6.7, and then $f(\rho^{e,-}, \mu^{e,-}) > f(\rho^e_{\max}, \mu^{e+1})$ and $\rho^{e+1,+} = \rho^{e+1}_{\max}$. The value $\mu^{e,+}$ is computed solving the equation $f(\rho^{e,-}, \mu^{e,+}) = f(\rho^{e+1}_{\max}, \mu^{e+1})$.

Observe that $f(\rho^{e,-}, \hat{\mu}^e) = \mu^{e,+} + \varepsilon(\rho^{e,-} - \mu^{e,+})$. In fact, supposing that $f(\rho^{e,-}, \hat{\mu}^e) = \rho^{e,-}$, since $\mu^{e,+} < \mu^{e,-}$ we get $f(\rho^{e,-}, \mu^{e,-}) = \rho^{e,-} = f(\rho^{e,-}, \mu^{e,+})$ but $f(\rho^{e,-}, \mu^{e,+}) > f(\rho^{e+1}_{\max}, \mu^{e+1})$, and there is no solution. We have

$$f(\rho^{e+1}_{\max}, \mu^{e+1,-}) - f(\rho^{e+1}_{\max}, \mu^{e+1}) = (1-\varepsilon)(\mu^{e+1,-} - \mu^{e+1}) = (1-\varepsilon)T.V.(\mu)^-.$$

Moreover, since

$$f(\rho^{e,-}, \mu^{e,-}) = f(\rho^{e+1,-}, \mu^{e+1,-}) \leq f(\rho^{e+1}_{\max}, \mu^{e+1,-}),$$

we have that

$$f(\rho^{e,-}, \mu^{e,-}) - f(\rho^{e,-}, \mu^{e,+}) \leq f(\rho^{e+1}_{\max}, \mu^{e+1,-}) - f(\rho^{e+1}_{\max}, \mu^{e+1}) = (1-\varepsilon)T.V.(\mu)^-.$$

Since we are in case (b) of Proposition 6.7, we have that

$$f(\rho^{e,-}, \mu^{e,-}) - f(\rho^{e,-}, \mu^{e,+}) = (1-\varepsilon)T.V.(\mu)^+ \leq (1-\varepsilon)T.V.(\mu)^-,$$

and then

$$T.V.(\mu)^+ \leq T.V.(\mu)^-. \quad \square$$

Estimates on Density Variation

We are ready to estimate the total variation of the function ρ.

Lemma 6.14. *Consider a wave-front tracking solution with initial data such that $T.V.(\mu(0)) = 0$. Then,*

$$T.V.(\rho(t)) \leq \frac{1}{\varepsilon} T.V.(\rho(0+)).$$

Proof. From Proposition 6.11, the chain rule for BV functions, we have that

$$T.V.(\rho(t,\cdot)) \le \frac{1}{\varepsilon} T.V.(f(t)) \le \frac{1}{\varepsilon} T.V.(f(0+)) \le \frac{1}{\varepsilon} T.V.(\rho(0+)). \quad \square$$

Thanks to Lemma 6.14 and the Lipschitz continuous dependence in L^1_{loc} of wave-front tracking approximations, we can apply the Helly theorem, as in [10], to get existence of solutions.

Theorem 6.15. *Fix a supply chain and assume* (F). *Given* $T > 0$, *for every initial datum such that* $T.V.(\mu(0)) = 0$, *there exists an admissible solution to the Cauchy problem on the network defined on* $[0,T]$.

Example of Unbounded Increase of $T.V.(\rho)$

We consider a simple chain formed just by two arcs $e_- =]-\infty, 0[$, $e_+ =]0, +\infty[$ connected at a vertex $v^e = 0$. On the chain the density ρ and the processing rate μ take values in the interval $[0, \rho_{\max}]$. As initial data we take

$$\rho_-(0) = \bar{\rho} + \frac{\delta}{2} \sum_{n=1}^{N} (1 + (-1)^n) \chi_{]-n,-n+1]}, \qquad \mu_-(0) = \rho_{\max},$$

$$\rho_+(0) = \bar{\rho}, \qquad \mu_+(0) = \bar{\rho},$$

where $\bar{\rho} \in]0, \rho_{\max}[$ and $\bar{\rho} + \delta < \rho_{\max}$. Notice that

$$T.V.(\rho(0)) = N\delta.$$

Consider the flux f_ε with $\varepsilon\Delta < \delta$, where $\Delta = \rho_{\max} - \bar{\rho}$. See Figure 6.11.

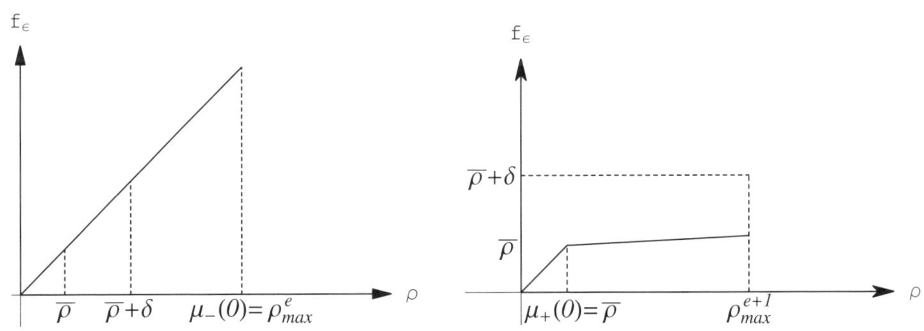

Figure 6.11. *Initial configuration.*

When the first ρ wave $(\bar{\rho} + \delta, \bar{\rho})$ reaches v^e, a μ wave $(\mu_-(0), \mu_-)$ is produced on e_- and a ρ wave $(\rho_{\max}, \bar{\rho})$ is produced on e_+, where

$$\mu_- = \frac{\bar{\rho}}{1-\varepsilon} - \frac{\varepsilon(\Delta - \bar{\rho} - \delta)}{1-\varepsilon}.$$

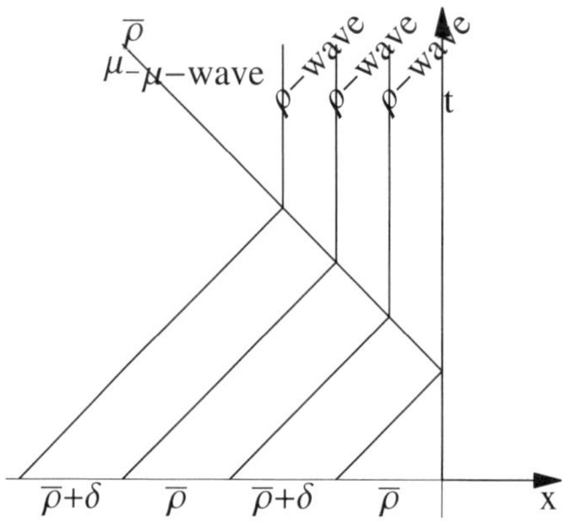

Figure 6.12. *Solution for $\varepsilon \to 0$.*

Notice that $\delta < \Delta$ implies $\mu_- > \bar{\rho}$ and μ_- tends to $\bar{\rho}$ as $\varepsilon \to 0$. After this interaction, the other waves from I_- produce a sequence of waves $(\rho_{\max}, \bar{\rho})$ and $(\bar{\rho}, \rho_{\max})$.

On the arc e_-, the μ wave interacts with the ρ waves with the following effect.

Case 1 The μ wave interacts with a wave $(\bar{\rho}, \bar{\rho} + \delta)$. Then the two waves cross each other and the velocity of the ρ wave after interaction is lowered to the value

$$\frac{(1-\varepsilon)(\mu_- - \bar{\rho}) + \varepsilon\delta}{\delta},$$

which tends to zero as $\varepsilon \to 0$.

Case 2 The μ wave interacts with a wave $(\bar{\rho} + \delta, \bar{\rho})$. Then the ρ wave is split into two waves $(\bar{\rho} + \delta, \mu_-)$ and $(\mu_-, \bar{\rho})$. The second wave travels with velocity equal to 1, while the first travels with velocity ε.

As a result at time $T = N + 1/2$ all ρ waves from e_- interacted with v^e; thus we get the estimates

$$T.V.(\rho(T)) = N\Delta = \frac{\Delta}{\delta} T.V.(\rho(0)),$$

and thus the magnification can be taken arbitrarily large choosing δ, and thus ε, sufficiently small.

As $\varepsilon \to 0$, the solution tends to the function $(\rho(t), \mu(t))$ represented in Figure 6.12.

6.7 A Riemann Solver According to Rule SC2

Rule SC2 individuates a unique Riemann Solver as shown by the next theorem.

Theorem 6.16. *Fix a node v^e. For every Riemann initial datum $(\rho^{e,0}, \mu^{e,0}, \rho^{e+1,0}, \mu^{e+1,0})$ at v^e there exists a unique vector $(\hat{\rho}^e, \hat{\mu}^e, \hat{\rho}^{e+1}, \hat{\mu}^{e+1})$ solution of the Riemann problem according to rule SC2.*

Proof. Given the initial datum $(\rho^{e,0}, \mu^{e,0}, \rho^{e+1,0,}, \mu^{e+1,0})$, it holds that

$$\hat{\rho}^e = \varphi(\hat{\mu}^e),$$
$$\hat{\mu}^{e+1} = \mu^{e+1,0},$$

where $\varphi(\hat{\mu}^e)$ has been defined in Section 6.4. We have to distinguish again two cases: Cases α and β.

Case α $\mu^{e+1,0} < \bar{\mu}^e$. We consider two subcases, Cases $\alpha_1 : \rho^* \le \rho_M(\mu^{e+1,0})$ and $\alpha_2 : \rho^* > \rho_M(\mu^{e+1,0})$, where

$$\rho^* = \frac{\bar{\mu}^e - (1-\varepsilon)\mu^{e+1,0}}{\varepsilon}. \tag{6.28}$$

Notice that the two cases correspond to the situation in which solutions in region B exist or do not exist.

Case α_1 Since $\mu^{e+1,0} < \bar{\mu}^e$ we get

$$\rho^* = \frac{1}{\varepsilon}\bar{\mu}^e - \left(\frac{1}{\varepsilon} - 1\right)\mu^{e+1,0} > \frac{1}{\varepsilon}\bar{\mu}^e - \left(\frac{1}{\varepsilon} - 1\right)\bar{\mu}^e = \bar{\mu}^e.$$

Consider the lines of Figure 6.7(right). To every μ there corresponds a value of the flux. We claim the following.

Claim If $\rho^* \le \rho_M$, the flux increases with respect to μ along the dashed lines in regions C, D and in region B for $\mu^e_{\max} \le \mu \le \rho^*$ and, finally, it is constant along the dashed line in region B for $\rho^* \le \mu \le \mu^e_{\max}$.

It holds that

$$f(\rho^*, \mu) = \begin{cases} \varepsilon\rho^* + (1-\varepsilon)\mu, & 0 \le \mu \le \rho^*, \\ \rho^*, & \rho^* \le \mu \le \mu^e_{\max}, \end{cases}$$

whose derivative, with respect to μ, is given by

$$\frac{\partial f}{\partial \mu}(\rho^*, \mu) = \begin{cases} 1-\varepsilon, & 0 \le \mu \le \rho^*, \\ 0, & \rho^* \le \mu \le \mu^e_{\max}. \end{cases}$$

It follows that for $\rho^* \le \hat{\mu}^e \le \mu^e_{\max}$ the flux is constant along the dashed line in region B. Let us now prove that the flux is increasing with respect to μ along the dashed lines in regions C and D. The line connecting the points $(\frac{2\varepsilon}{1+\varepsilon}\bar{\mu}^e, 0)$ and $(\mu^{e+1,0}, \mu^*)$ with $\mu^* = \frac{1+\varepsilon}{1-\varepsilon}(\hat{\mu}^{e+1} - \frac{2\varepsilon}{1+\varepsilon}\bar{\mu}^e)$ has equation

$$\rho - \frac{1}{\mu^*}\left(\mu^{e+1,0} - \frac{2\varepsilon}{1+\varepsilon}\bar{\mu}^e\right)\mu - \frac{2\varepsilon}{1+\varepsilon}\bar{\mu}^e = 0,$$

and a directional vector is given by

$$r_{C_\alpha} = \begin{pmatrix} \frac{1}{\mu^*}\left(\mu^{e+1,0} - \frac{2\varepsilon}{1+\varepsilon}\bar{\mu}^e\right) \\ 1 \end{pmatrix}.$$

Therefore, the directional derivative of the flux is equal to

$$\nabla f(\rho,\mu)\cdot r_{C_\alpha} = \begin{pmatrix} \varepsilon \\ 1-\varepsilon \end{pmatrix} \cdot \begin{pmatrix} \frac{1}{\mu^*}\left(\mu^{e+1,0} - \frac{2\varepsilon}{1+\varepsilon}\bar{\mu}^e\right) \\ 1 \end{pmatrix}$$

$$= \frac{\varepsilon}{\mu^*}\left(\mu^{e+1,0} - \frac{2\varepsilon}{1+\varepsilon}\bar{\mu}^e\right) + (1-\varepsilon) > 0.$$

The latter inequality is fulfilled if $\mu^{e+1,0} > \frac{2\varepsilon}{1+\varepsilon}\bar{\mu}^e$, which is true whenever we have solutions in region C. In region D a directional vector of the line connecting the points $(\mu^{e+1,0}, \mu^*)$ and $(\rho^*, \bar{\mu}^e)$ is the following:

$$r_{D_\alpha} = \begin{pmatrix} \frac{\rho^* - \mu^{e+1,0}}{\bar{\mu}^e - \mu^*} \\ 1 \end{pmatrix}.$$

It implies that

$$\nabla f(\rho,\mu)\cdot r_{D_\alpha} = \begin{pmatrix} \varepsilon \\ 1-\varepsilon \end{pmatrix} \cdot \begin{pmatrix} \frac{\rho^* - \mu^{e+1,0}}{\bar{\mu}^e - \mu^*} \\ 1 \end{pmatrix} = \varepsilon\frac{\rho^* - \mu^{e+1,0}}{\bar{\mu}^e - \mu^*} + (1-\varepsilon) > 0,$$

since $\rho^* > \bar{\mu}^e > \mu^{e+1,0}$ and $\bar{\mu}^e - \mu^* > 0$. In order to respect rule SC2 we set

$$\hat{\rho}^{e+1} = \rho^*, \hat{\mu}^e = \min\{\mu^e_{\max}, \rho^*\}.$$

Case α_2 If $\rho^* > \rho_M(\mu^{e+1,0})$, there are no solutions in region B, and since the flux increases with respect to μ in region D we set

$$\hat{\rho}^{e+1} = \rho_M, \quad \hat{\mu}^e = \tilde{\mu},$$

where $\tilde{\mu}$ is obtained from

$$(1-\varepsilon)\mu^{e+1,0} + \varepsilon\hat{\rho}^{e+1} = (1-\varepsilon)\hat{\mu}^e + \varepsilon\hat{\rho}^e,$$

setting $\hat{\rho}^{e+1} = \rho_M$, i.e.,

$$\tilde{\mu} = \frac{\varepsilon(1+\varepsilon)}{1-\varepsilon}\rho_M - \frac{2\varepsilon}{1-\varepsilon}\bar{\mu}^e + (1+\varepsilon)\mu^{e+1,0} = \frac{2\varepsilon}{1-\varepsilon}(\mu^e_{\max} - \bar{\mu}^e) + \mu^{e+1,0}.$$

Case β $\bar{\mu}^e \le \mu^{e+1,0}$. Consider the line of Figure 6.7(left). In this case the flux is constant with respect to μ along the line in region A and is an increasing function along the line in region C. In fact, since the line in region A is given by $\hat{\rho}^{e+1} = \bar{\mu}^e$, it follows that

$$f\left(\hat{\rho}^{e+1},\mu\right) = \begin{cases} \varepsilon\bar{\mu}^e + (1-\varepsilon)\mu, & 0 \le \mu \le \bar{\mu}^e, \\ \rho, & \bar{\mu}^e \le \mu \le \mu^e_{\max}, \end{cases}$$

from which

$$\frac{\partial f}{\partial \mu}\left(\hat{\rho}^{e+1},\mu\right)=\begin{cases}(1-\varepsilon), & 0\le\mu\le\bar{\mu}^e,\\ 0, & \bar{\mu}^e\le\mu\le\mu^e_{\max}.\end{cases}$$

In region C the line connecting the points $(\frac{2\varepsilon}{1+\varepsilon}\bar{\mu}^e,0)$ and $(\bar{\mu}^e,\bar{\mu}^e)$ has equation

$$\rho-\frac{1-\varepsilon}{1+\varepsilon}\mu-\frac{2\varepsilon}{1+\varepsilon}\bar{\mu}^e=0,$$

and a directional vector is given by

$$r_{C_\beta}=\begin{pmatrix}\frac{1-\varepsilon}{1+\varepsilon}\\ 1\end{pmatrix}.$$

The directional derivative is the following:

$$\nabla f(\rho,\mu)\cdot r_{C_\beta}=\begin{pmatrix}\varepsilon\\ 1-\varepsilon\end{pmatrix}\cdot\begin{pmatrix}\frac{1-\varepsilon}{1+\varepsilon}\\ 1\end{pmatrix}=\varepsilon\frac{1-\varepsilon}{1+\varepsilon}+(1-\varepsilon)>0.$$

It follows that rule SC2 is satisfied if we define

$$\hat{\rho}^{e+1}=\bar{\mu}^e,\quad \hat{\mu}^e=\bar{\mu}^e.$$

Finally the Riemann Solver is the following.

Case α $\mu^{e+1,0}<\bar{\mu}^e$

Case α_1 $\rho^*\le\rho_M(\mu^{e+1,0})$

$$\hat{\rho}^{e+1}=\rho^*,\quad \hat{\mu}^e=\min\{\mu^e_{\max},\rho^*\}.$$

Case α_2 $\rho^*>\rho_M(\mu^{e+1,0})$

$$\hat{\rho}^{e+1}=\rho_M(\mu^{e+1,0}),\quad \hat{\mu}^e=\tilde{\mu}.$$

Case β $\mu^{e+1,0}\ge\bar{\mu}^e$

$$\hat{\rho}^{e+1}=\bar{\mu}^e,\quad \hat{\mu}^e=\bar{\mu}^e.\quad\square$$

This Riemann Solver is depicted in Figure 6.13. In Case β we can define a Riemann Solver mapping every initial datum to the point $(\bar{\mu}^e,\bar{\mu}^e)$, indicated by the arrow.

In Case α we can define a Riemann Solver mapping every initial datum to the circle or to the square point if $\rho^*\le\rho_M$ and to the filled point if $\rho^*>\rho_M$.

6.8 A Riemann Solver According to Rule SC3

Also rule SC3 determines a unique Riemann Solver as shown by the next theorem.

Theorem 6.17. *Fix a node v^e. For every Riemann initial datum $(\rho^{e,0},\mu^{e,0},\rho^{e+1,0},\mu^{e+1,0})$ at v^e there exists a unique vector $(\hat{\rho}^e,\hat{\mu}^e,\hat{\rho}^{e+1},\hat{\mu}^{e+1})$ solution of the Riemann problem according to rule SC3.*

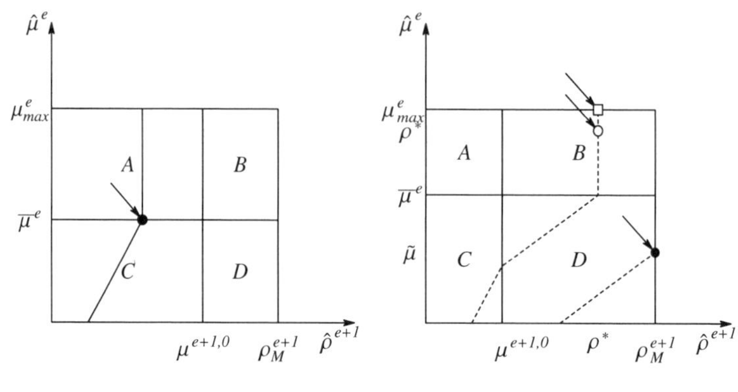

Figure 6.13. *Case β (on the left) and Case α (on the right) for the Riemann solver SC2.*

Proof. As for the Riemann Solver for rule SC2, given the initial datum $(\rho^{e,0}, \mu^{e,0}, \rho^{e+1,0}, \mu^{e+1,0})$, we have

$$\hat{\rho}^e = \varphi(\hat{\mu}^e),$$
$$\hat{\mu}^{e+1} = \mu^{e+1,0}.$$

Cases α and β are distinguished.

Case α This case is further split into subcases. First we have Cases α_1 and α_2, depending on which between ρ^* and $\rho_M(\mu^{e+1,0})$ is greater.

Case α_1 $\rho^* \le \rho_M(\mu^{e+1,0})$. In Theorem 6.16 it was proved that the flux increases with respect to μ along the dashed lines in regions C, D and in region B for $\mu^e_{\max} \le \mu \le \rho^*$ and, finally, that it is constant along the line in region B for $\rho^* \le \mu \le \mu^e_{\max}$. It follows that we have to consider two possible situations: Cases $\alpha_{1.1} : \rho^* > \mu^e_{\max}$ and $\alpha_{1.2} : \rho^* \le \mu^e_{\max}$.

Case $\alpha_{1.1}$ $\rho^* > \mu^e_{\max}$. According to rule SC3 we set

$$\hat{\rho}^{e+1} = \rho^*, \quad \hat{\mu}^e = \mu^e_{\max}.$$

Case $\alpha_{1.2}$ $\rho^* \le \mu^e_{\max}$. We set

$$\hat{\rho}^{e+1} = \rho^*, \quad \hat{\mu}^e = \max\{\rho^*, \mu^{e,0}\}.$$

Case α_2 $\rho^* > \rho_M(\mu^{e+1,0})$. In this case, there are not solutions in region B, and since the flux increases with respect to μ in region D we set, as for the Riemann Solver SC2,

$$\hat{\rho}^{e+1} = \rho_M(\mu^{e+1,0}), \quad \hat{\mu}^e = \tilde{\mu}.$$

Case β The flux is constant with respect to μ along the line in region A and is an increasing function along the line in region C; then we set

$$\hat{\rho}^{e+1} = \bar{\mu}^e, \; \hat{\mu}_e = \begin{cases} \bar{\mu}^e, & \text{if } \mu^{e,0} < \bar{\mu}^e, \\ \mu^{e,0}, & \text{if } \mu^{e,0} \ge \bar{\mu}^e. \end{cases} \qquad \square$$

The obtained Riemann Solver is depicted in Figure 6.14: all points of the white region are mapped horizontally and all points of the dark regions are mapped to the point indicated by the arrows.

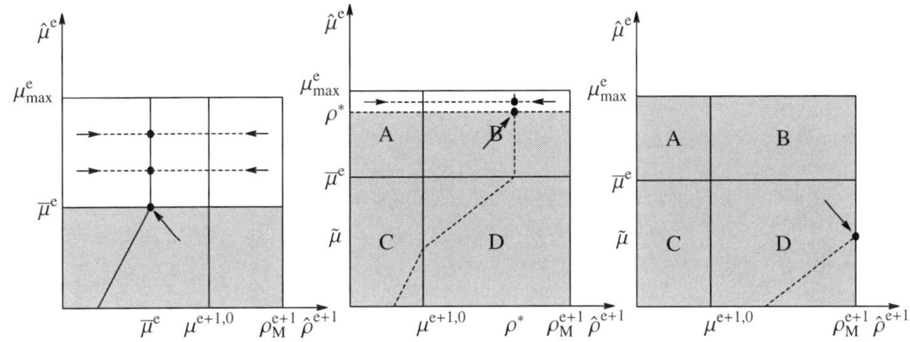

Figure 6.14. *Cases β and α (namely Cases α_1 and α_2) for the Riemann Solver SC3.*

Analogously to the case of rule SC1, we can give conditions for the solvability of Riemann problems; more precisely, we have the following lemma.

Lemma 6.18. *Consider a supply chain on which the initial datum verifies $\mu^{e,0} = \mu^e_{max}$; i.e., the production rate is at its maximum. A sufficient condition for the solvability of all Riemann problems, according to rule SC2 or SC3, on the supply chain at every time is*

$$\rho^{e+2}_{max} \geq \rho^e_{max} \quad \forall e.$$

The proof is similar to Lemma 6.9 for rule SC1.

6.9 Analysis of Equilibria

In this section we discuss the equilibria at nodes. We fix a node v^e and a Riemann initial datum $\rho_0 = (\rho^{e,0}, \mu^{e,0}, \rho^{e+1,0}, \mu^{e+1,0})$.

Definition 6.19. *Define $\hat{\rho} = RS(\rho_0)$. The datum ρ_0 is an equilibrium if*

$$\hat{\rho} = RS(\rho_0) = \rho_0.$$

The conservation of flux at the node can be written as

$$f(\rho^{e,0}, \mu^{e,0}) = f(\rho^{e+1,0}, \mu^{e+1,0}),$$

and it is satisfied in accordance with the Riemann Solvers. From now on, we suppose that the arcs have the same maximum processing rate, i.e, $\mu^e_{max} = \mu^{e+1}_{max}$ for all e.

Riemann Solver SC1

Let us analyze Cases $\alpha : \mu^{e+1,0} < \bar{\mu}^e$ and $\beta : \mu^{e+1,0} \geq \bar{\mu}^e$.

Case α We have to discuss the subcases, Cases $\alpha_1 : \rho^* \leq \rho_M(\mu^{e+1,0})$ and $\alpha_2 : \rho^* > \rho_M(\mu^{e+1,0})$.

Case α_1 In this case $\mu^{e,0}, \rho^{e,0}$, and $\mu^{e+1,0}$ can assume all positive values, with $\mu^{e+1,0}$ subject to the constraint $\mu^{e+1,0} < \bar{\mu}^e = \bar{\mu}^e(\rho^{e,0}, \mu^{e,0})$, while $\rho^{e+1,0} = \rho^{e+1,0}(\rho^{e,0}, \mu^{e+1,0})$.

Case α_2 We have $\mu^{e,0} \leq \tilde{\mu}(\rho^{e,0}, \mu^{e+1,0})$. Since $\mu^{e,0} < \bar{\mu}^e$, it follows that $\rho^{e,0} \geq \mu^{e,0}$, from which we obtain $\bar{\mu}^e = \mu^{e,0} + \varepsilon(\rho^{e,0} - \mu^{e,0})$. Finally, we have that $\rho^{e,0}$, $\mu^{e,0}$, and $\mu^{e+1,0}$ can assume all the positive values, with $\mu^{e+1,0}$ subject to the conditions $\mu^{e+1,0} \leq \mu^{e,0} + \varepsilon(\rho^{e,0} - \mu^{e,0}), \mu^{e,0} \leq \tilde{\mu}(\rho^{e,0}, \mu^{e+1,0})$, and $\rho^{e+1,0} = \rho^{e+1,0}(\rho^{e,0}, \mu^{e+1,0})$.

Case β In this case $\rho^{e,0}$, $\mu^{e,0}$, and $\mu^{e+1,0}$ can assume all positive values, with $\mu^{e+1,0}$ subject to the constraint $\mu^{e+1,0} \geq \bar{\mu}^e$ and $\rho^{e+1,0} = \rho^{e+1,0}(\rho^{e,0}, \mu^{e+1,0})$.

Riemann Solver SC2

Let us distinguish Cases α and β.

Case α We have to consider the subcases, Cases $\alpha_1 : \rho^* < \rho_M(\mu^{e+1,0})$ and $\alpha_2 : \rho^* \geq \rho_M(\mu^{e+1,0})$. Let us start by considering Case α_1.

Case α_1 In this case, from the Riemann Solver we get

$$\rho^{e+1,0} = \rho^* > \bar{\mu}^e, \quad \mu^{e,0} = \min\{\mu^e_{\max}, \rho^*\} \leq \rho^{e+1,0}, \quad \mu^{e,0} \geq \bar{\mu}^e.$$

From the latter inequality, using (6.15), it follows that $\bar{\mu}^e = \rho^{e,0}$. The hypothesis $\mu^{e+1,0} < \bar{\mu}^e$ implies that

$$\mu^{e+1,0} < \bar{\mu}^e = \rho^{e,0} \leq \mu^{e,0} \leq \rho^{e+1,0}. \tag{6.29}$$

Moreover, the relation $\rho^* < \rho_M(\mu^{e+1,0})$ leads to

$$\rho^{e,0} < \frac{2\varepsilon}{1+\varepsilon}\mu^e_{\max} + \frac{1-\varepsilon}{1+\varepsilon}\mu^{e+1,0}. \tag{6.30}$$

Remark 6.9.1. *If $\varepsilon = 0$, from* (6.30) *we get $\rho^{e,0} < \mu^{e+1,0}$, a relation that is in contradiction with* (6.29). *In fact in this case ρ^* is not defined. It follows that there are not equilibria for $\varepsilon = 0$.*

Since $\mu^{e,0} = \min\{\mu^e_{\max}, \rho^*\}$, consider now Cases $\alpha_{1.1} : \mu^e_{\max} \leq \rho^*$ and $\alpha_{1.2} : \mu^e_{\max} > \rho^*$.

Case $\alpha_{1.1}$ In this case we have

$$\begin{aligned} \mu^{e,0} &= \mu^e_{\max}, \\ \rho^{e+1,0} &\geq \mu^e_{\max}. \end{aligned} \tag{6.31}$$

Taking into account that $\rho^{e+1} = \rho^*$ and using (6.28) and (6.30) we get

$$\mu^e_{\max} \leq \rho^{e+1,0} < \frac{2}{1+\varepsilon}\mu^e_{\max} - \frac{1-\varepsilon}{1+\varepsilon}\mu^{e+1,0} = \rho_M(\mu^{e+1,0}). \tag{6.32}$$

Starting from a fixed value of $\rho^{e,0}$, we discuss the equilibria. Since (6.30) holds and supposing that $\rho^{e,0}$ assumes the maximum value

$$\frac{2\varepsilon}{1+\varepsilon}\mu^e_{\max} + \frac{1-\varepsilon}{1+\varepsilon}\mu^{e+1,0},$$

we get the following value for $\mu^{e+1,0}$:

$$\mu_1(\rho^{e,0}) = \frac{1+\varepsilon}{1-\varepsilon}\rho^{e,0} - \frac{2\varepsilon}{1-\varepsilon}\mu^e_{\max}. \tag{6.33}$$

From (6.29) and (6.31) it follows that if $\rho^{e,0} \in [0, \mu^e_{\max}]$, then $\mu^{e+1,0} \in [\mu_1, \rho^{e,0}]$. The relation (6.32) implies that the maximum value $\rho^{e+1,0}$ can assume is given by

$$\rho^{e+1,0} = \rho^{e+1,0}(\rho^{e,0}, \mu^{e+1,0}) = \frac{2\varepsilon}{1+\varepsilon}\mu^e_{\max} - \frac{1-\varepsilon}{1+\varepsilon}\mu^{e+1,0},$$

which is a decreasing function with respect to $\mu_{e+1,0}$.
Using (6.33) we can express $\rho^{e+1,0}$ as a function of $\rho^{e,0}$ and $\mu^e_{\max}$:

$$\rho^{e+1,0} = \rho^* = 2\mu^e_{\max} - \rho^{e,0};$$

therefore $\rho^{e+1,0} \in [\mu^e_{\max}, 2\mu^e_{\max} - \rho^{e,0}]$. Since $\rho^{e+1,0} \geq \mu^e_{\max}$, it follows that $\mu^{e+1,0} \in [\mu_1(\rho^{e,0}), \mu_2(\rho^{e,0})]$, where

$$\mu_2(\rho^{e,0}) = \frac{1}{1-\varepsilon}\rho^{e,0} - \frac{\varepsilon}{1-\varepsilon}\mu^e_{\max}.$$

It remains to establish the conditions ensuring that $\mu_2(\rho^{e,0}) \geq 0$ and $\mu_2(\rho^{e,0}) \leq \mu^e_{\max}$. The second relation is always fulfilled since $\rho^{e,0} \leq \mu^{e,0}$. Instead, the first holds if

$$\rho^{e,0} \geq \varepsilon\mu^{e,0} = \varepsilon\mu^e_{\max}. \tag{6.34}$$

Remark 6.9.2. *Notice that if $\rho^{e,0} = \varepsilon\mu^{e,0}$, we have $\mu^{e+1,0} = 0$ and $\rho^{e+1,0} = \mu^e_{\max}$.*

If (6.33) is satisfied, the equilibria configurations depend on the value of $\mu_1(\rho^{e,0})$. We have that $\mu_1(\rho^{e,0}) \gtreqless 0$ if $\rho^{e,0} \gtreqless \frac{2\varepsilon}{1+\varepsilon}\mu^e_{\max}$. In the case $\mu_1(\rho^{e,0}) < 0$, since $\mu^{e+1,0} = 0$ implies that $\rho^{e+1,0} = \frac{\rho^{e,0}}{\varepsilon}$, we obtain that $\rho^{e+1,0} \in [\mu^e_{\max}, \frac{\rho^{e,0}}{\varepsilon}]$.

Finally, the set of equilibria is defined in the following way: $\mu^{e,0} = \mu^e_{\max}$, $\rho^{e,0} \in [\varepsilon\mu^e_{\max}, \mu^e_{\max}]$. In particular, we have to consider two cases.

Case $\alpha_{1.1.1}$ If $\rho^{e,0} \in [\varepsilon\mu^e_{\max}, \frac{2\varepsilon}{1+\varepsilon}\mu^e_{\max}]$, then we have $\mu^{e+1,0} \in [0, \mu_2(\rho^{e,0})]$, $\rho^{e+1,0} = \rho^*(\rho^{e,0}, \mu^{e+1,0}) \in [\mu^e_{\max}, \frac{\rho^{e,0}}{\varepsilon}]$.

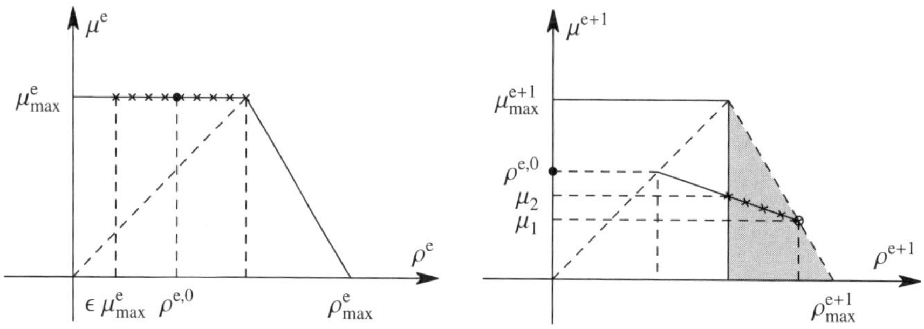

Figure 6.15. *Equilibria configurations for Case* $\alpha_{1.1}$.

Case $\alpha_{1.1.2}$ If $\rho^{e,0} \in [\frac{2\varepsilon}{1+\varepsilon}\mu^e_{\max}, \mu^e_{\max}]$, then we have $\mu^{e+1,0} \in [\mu_1(\rho^{e,0}), \mu_2(\rho^{e,0})]$, $\rho^{e+1,0} = \rho^*(\rho^{e,0}, \mu^{e+1,0}) \in [\mu^e_{\max}, \rho_M(\mu^{e+1,0})]$.

Equilibria configurations are depicted in Figure 6.15. Observe that for fixed $\rho^{e,0}$ we have a segment on the outgoing arc. If $\rho^{e,0}$ varies in the interval $[\varepsilon\mu^e_{\max}, \mu^e_{\max}]$, we get a region on the outgoing arc, colored in gray, without the oblique side. Moreover, the processing rate is maximum on the incoming arc and assumes values in $[\mu_1(\rho^{e,0}), \mu_2(\rho^{e,0})]$ on the incoming one, while the density on the outgoing arc is greater than the density on the incoming one.

Case $\alpha_{1.2}$ From $\mu^e_{\max} > \rho^*$ it follows that $\rho^{e+1,0} = \mu^{e,0} = \rho^*$. Finally we have

$$\mu^{e+1,0} < \rho^{e,0} \leq \rho^{e+1,0} = \rho^*(\rho^{e,0}, \mu^{e+1,0}) < \mu^e_{\max}.$$

In this case $\rho^* = \rho^{e+1,0} \in [\rho^{e,0}, \mu^e_{\max}]$; in fact if $\mu^{e+1,0} = \rho^{e,0}$, we have $\rho^* = \rho^{e,0}$. From Case $\alpha_{1.1}$ we get $\rho^{e,0} \in [\varepsilon\mu^e_{\max}, \mu^e_{\max}]$. We have to distinguish two cases.

Case $\alpha_{1.2.1}$ If $\rho^{e,0} \in [0, \varepsilon\mu^e_{\max}]$, then $\mu^{e+1,0} \in [0, \rho^{e,0}]$ and $\rho^{e+1,0} = \rho^*(\rho^{e,0}, \mu^{e+1,0}) = \mu^{e,0} \in [\rho^{e,0}, \frac{\rho^{e,0}}{\varepsilon}]$.

Case $\alpha_{1.2.2}$ If $\rho^{e,0} \in [\varepsilon\mu^e_{\max}, \mu^e_{\max}]$, then we have $\mu^{e+1,0} \in [\mu_2(\rho^{e,0}), \rho^{e,0}]$ and $\rho^{e+1,0} = \rho^*(\rho^{e,0}, \mu^{e+1,0}) = \mu^{e,0} \in [\rho^{e,0}, \mu^e_{\max}]$.

Equilibria configurations are depicted in Figure 6.16. Observe that at a point on the incoming arc corresponds a point on the outgoing arc. Moreover, the density on the outgoing arc is equal to the processing rate on the incoming arc.

Case α_2 In this case $\mu^{e+1,0} < \bar{\mu}^e$. According to the Riemann Solver

$$\mu^{e,0} = \tilde{\mu} = \frac{2\varepsilon}{1-\varepsilon}(\mu^e_{\max} - \bar{\mu}^e) + \mu^{e+1,0}. \tag{6.35}$$

Since $\mu^{e,0} = \tilde{\mu} \leq \bar{\mu}^e$, from the definition of $\bar{\mu}^e$ it follows that

$$\bar{\mu}^e = \frac{1+\varepsilon}{2}\rho^{e,0} + \frac{1-\varepsilon}{2}\mu^{e,0}, \tag{6.36}$$

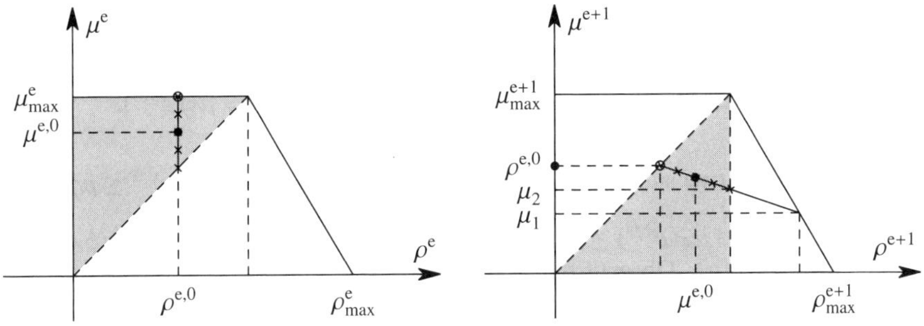

Figure 6.16. *Equilibria configurations for Case $\alpha_{1.2}$.*

and $\rho^{e,0} \geq \mu^{e,0}$. Using (6.36), the relation (6.35) reads

$$\mu^{e,0} = \frac{2\varepsilon}{1-\varepsilon^2}\mu^e_{max} - \frac{\varepsilon}{1-\varepsilon}\rho^{e,0} + \frac{1}{1+\varepsilon}\mu^{e+1,0}. \tag{6.37}$$

Finally, we have

$$\begin{aligned} \rho^{e,0} \geq \mu^{e,0} &= \frac{2\varepsilon}{1-\varepsilon^2}\mu^e_{max} - \frac{\varepsilon}{1-\varepsilon}\rho^{e,0} + \frac{1}{1+\varepsilon}\mu^{e+1,0}, \\ \mu^{e+1,0} &\leq \frac{1+\varepsilon}{2}\rho^{e,0} + \frac{1-\varepsilon}{2}\mu^{e,0}, \\ \rho^{e+1,0} &= \rho_M(\mu^{e+1,0}). \end{aligned} \tag{6.38}$$

After straightforward computations, substituting (6.37) in (6.38) we have

$$\mu^{e+1,0} \leq \frac{2\varepsilon}{1+3\varepsilon}\mu^e_{max} + \frac{1+\varepsilon}{1+3\varepsilon}\rho^{e,0} = \mu_3(\rho^{e,0}). \tag{6.39}$$

From $\rho^* \geq \rho_M(\mu^{e+1,0})$ or $\mu^{e,0} \leq \rho^{e,0}$, it follows that

$$\mu^{e+1,0} \leq -\frac{2\varepsilon}{1-\varepsilon}\mu^e_{max} + \frac{1+\varepsilon}{1-\varepsilon}\rho^{e,0} = \mu_4(\rho^{e,0}). \tag{6.40}$$

Notice that $\mu_3(\rho^{e,0}) \geq \mu_4(\rho^{e,0})$ if $\rho^{e,0} \leq \mu^e_{max}$. Therefore if the latter holds, (6.39) is fulfilled if (6.40) is true.

In order to have equilibria configurations, the following relations must hold: $\mu^{e,0} \geq 0$ and $\mu^{e,0} \leq \mu^e_{max}$. From the condition $\mu^{e,0} \geq 0$, we obtain

$$\rho^{e,0} \geq 2\mu^e_{max} + \frac{1-\varepsilon}{\varepsilon(1+\varepsilon)}\mu^{e+1,0},$$

which is always true since $\rho^e_{max} = \frac{2}{1+\varepsilon}\mu^e_{max}$.

Equilibria for $\varepsilon = 0$.

We compare the equilibria obtained using the rules SC1, SC2, and SC3 in the case $\varepsilon = 0$. This permits us to understand the typical feature of the supply chain at equilibria in the three cases.

$\varepsilon = 0$	SC1	SC2	SC3
α_1	No equilibria	No equilibria	No equilibria
α_2	$\mu^{e,0} = \mu^{e+1,0} \leq \rho^{e,0}$, $\rho^{e+1,0} = 2\mu_{\max}^{e+1} - \mu^{e+1,0}$,	Same as SC1	Same as SC1
β	either $\rho^{e,0} = \rho^{e+1,0} \leq \mu^{e,0}, \mu^{e+1,0}$, or $\rho^{e+1,0} = \mu^{e,0} \leq \rho^{e,0}$, $\mu^{e+1,0} \geq \frac{1}{2}\rho^{e,0} + \frac{1}{2}\mu^{e,0}$	$\rho^{e,0} = \rho^{e+1,0} = \mu^{e,0} \leq \mu^{e+1,0}$	First case of SC1

We first notice that the complicated equilibria of Case α_1 disappear from all Riemann Solvers.

Regarding Case α_2, all rules perform again in the same way. In particular, the processing rates are the same in the entering and exiting lines, while the entering density is greater. To keep the flux equilibrium the exiting density is settled accordingly (the precise value is obtained passing to the limit in $\varepsilon \to 0$).

Let us compare such equilibria with the model introduced in Section 5.2 of Chapter 5, which, for brevity, we call the Q model (model with queues). This case would correspond to a queue which is increasing in size, since the incoming density is bigger than the processing rate. Also to have this situation, we would need $\mu^{e+1,0} = \mu_{\max}^{e+1}$. Then we would also have $\rho^{e+1,0} = \mu_{\max}^{e+1}$, and the two models fit well.

Finally, we reproduce the situation of an increasing queue, but we may have some equilibria not possible for the model.

Let us now analyze Case β. The first case of SC1 (and thus the only of SC3) corresponds to equal densities, which are lower than the respective processing rates. For the Q model, this corresponds to an empty queue situation, where the incoming and outgoing fluxes are less than the production rates.

For SC2, we have a quite special case. The incoming density equals the processing rate, while the outgoing density is lower than the corresponding processing rate. Thus again, for the Q model we have an empty queue situation.

Finally, the second case of SC1 is quite different. In fact, we have the incoming density bigger than the processing rate, while, in the outgoing arc, the density, equal to the incoming one, is definitely below the processing rate. Again, this case corresponds to an empty queue.

Concluding, the model proposed in this chapter and that of Q are comparable for what concerns equilibria. As expected, the emptying queue situation does not appear for the present model, while the choice SC3 seems the more appropriate to reproduce the Q model features.

6.10 A Continuum-Discrete Model for Supply Networks

The aim of this section is to extend the continuum-discrete model regarding sequential supply chains to supply networks which consist of arcs and two types of nodes: nodes with one incoming arc and more outgoing ones and nodes with more incoming arcs and one outgoing arc. In fact, these two types of nodes are the most common in real supply networks.

Model Description

Let us introduce briefly the model.

Definition 6.20 (Network definition). *A supply network is a finite, connected directed graph consisting of a finite set of arcs $\mathcal{A} = \{1, \ldots, N+1\}$ and a finite set of junctions (vertices, nodes) $\mathcal{V}$.*

On each arc the load dynamic is given by a continuum system of type

$$\rho_t + f_\varepsilon^e(\rho, \mu)_x = 0, \tag{6.43}$$

$$\mu_t - \mu_x = 0, \tag{6.44}$$

where ρ is the density of objects processed by the supply chain network, μ is the processing rate, and f_ε^e is the flux given by

$$f_\varepsilon^e(\rho, \mu) = \begin{cases} \rho, & 0 \leq \rho \leq \mu, \\ \mu + \varepsilon(\rho - \mu), & \mu \leq \rho \leq \rho_{\max}^e, \end{cases}$$

with $\rho_{\max}^e$ and $\mu_{\max}^e$ the maximum density and processing rate on the arc e.

We interpret the evolution at a node v^e thinking of it as a Riemann problem for the density equation (6.43) with μ data as parameters. The Riemann problems are solved fixing two "routing" algorithms.

RA1 Goods from an incoming arc are sent to outgoing ones according to their final destination in order to maximize the flux over incoming arcs. Goods are processed, ordered by arrival time (FIFO policy).

RA2 Goods are processed by arrival time (FIFO policy) and are sent to outgoing arcs in order to maximize the flux over incoming and outgoing arcs.

The two algorithms were already used in [25] for the analysis of packet flows in telecommunication networks. Notice that the second algorithm allows the redirection of goods, taking into account possible high loads of outgoing arcs. For both routing algorithms the flux of goods is maximized considering one of the two additional rules, SC2 and SC3.

Riemann Solvers for Nodes

In this section we discuss Riemann Solvers, which conserve the flux at nodes. We consider two kinds of nodes:

a node with more incoming arcs and one outgoing one;

a node with one incoming arc and more outgoing ones.

We consider a node v^e with n incoming arcs and m outgoing ones and a Riemann initial datum $(\rho^{1,0},\mu^{1,0},\dots,\rho^{n,0},\mu^{n,0},\rho^{n+1,0},\mu^{n+1,0},\dots,\rho^{n+m,0},\mu^{n+m,0})$.

The following lemma holds.

Lemma 6.21. *On the incoming arcs, only waves of the first family may be produced, while on the outgoing arcs only waves of the second family may be produced.*

From Lemma 6.21, given the initial datum, for every Riemann Solver it follows that

$$\begin{aligned} \hat{\rho}^e &= \varphi(\hat{\mu}^e), \quad e \in \delta_v^-, \\ \hat{\mu}^e &= \mu^{e,0}, \quad\;\;\; e \in \delta_v^+, \end{aligned} \tag{6.45}$$

where the function $\varphi(\cdot)$ describes the first family curve through $(\rho^{e,0},\mu^{e,0})$ as a function of $\hat{\mu}^e$. The expression of such a curve changes at a particular value $\bar{\mu}^e$, given by

$$\bar{\mu}^e = \begin{cases} \rho^{e,0}, & \text{if } \rho^{e,0} \le \mu^{e,0}, \\ \frac{1+\varepsilon}{2}\rho^{e,0} + \frac{1-\varepsilon}{2}\mu^{e,0}, & \text{if } \rho^{e,0} > \mu^{e,0}. \end{cases}$$

We define two different Riemann Solvers at a junction that represent two different routing algorithms.

RA1 We assume the following:

(A) the flow from incoming arcs is distributed on outgoing arcs according to fixed coefficients;

(B) respecting (A) the processor chooses to process goods in order to maximize fluxes (i.e., the number of goods which are processed).

RA2 We assume that the number of goods through the junction is maximized over both incoming and outgoing arcs.

For both routing algorithms we can maximize the flux of goods considering one of the two additional rules SC2 and SC3.

To define Riemann problems according to algorithms RA1 and RA2 let us introduce the notation

$$f^e = f(\rho^e,\mu^e).$$

Define the maximum flux that can be obtained by a wave solution on each production arc:

$$f^e_{\max} = \begin{cases} \bar{\mu}^e, & e \in \delta_v^-, \\ \mu^{e,0} + \varepsilon(\rho_M(\mu^{e,0}) - \mu^{e,0}), & e \in \delta_v^+. \end{cases}$$

Since $\hat{f}^e \in [f^e_{\min}, f^e_{\max} = \bar{\mu}^e]$, $e \in \delta_v^-$, and $\hat{f}^e \in [0, f^e_{\max} = \mu^{e,0} + \varepsilon(\rho_M(\mu^{e,0}) - \mu^{e,0})]$, $e \in \delta_v^+$, it follows that if

$$\sum_{e\in\delta_v^-} f^e_{\min} > \sum_{e\in\delta_v^+} f^e_{\max},$$

the Riemann problem does not admit solution. Thus we get the following condition for the solvability of the supply network.

Lemma 6.22. *A necessary and sufficient condition for the solvability of the Riemann problems is that*

$$\sum_{e\in\delta_v^-} f_{\min}^e \le \sum_{e\in\delta_v^+} \mu^{e,o} + \varepsilon(\rho^M(\mu^{e,o}) - \mu^{e,o}).$$

Lemma 6.23. *A sufficient condition for the solvability of the Riemann problem, independent of the initial data, is the following:*

$$\sum_{e\in\delta_v^-} \rho_{\max}^e \le \sum_{e\in\delta_v^+} \mu_{\max}^e.$$

Proof. Since $\hat{f}^e \in [f_{\min}^e, f_{\max}^e]$, $e \in \delta_v^-$, and $\hat{f}^e \in [0, f_{\max}^e]$, $e \in \delta_v^+$, the worst case to fulfill the condition of Lemma 6.22 happens when $f_{\min}^e$, $e \in \delta_v^-$, assumes the greatest value and $f_{\max}^e$, $e \in \delta_v^+$, the lowest one:

$$\sum_{e\in\delta_v^-} \varepsilon\rho_{\max}^e \le \sum_{e\in\delta_v^+} \varepsilon\mu_{\max}^e. \quad \square$$

In what follows, first we consider a single junction $v^e \in \mathcal{V}$ with $n-1$ incoming arcs and 1 outgoing arc (shortly, a node of type $(n-1)\times 1$) and then a junction with 1 incoming arc and $m-1$ outgoing ones (shortly, a node of type $1\times(m-1)$).

One Outgoing Arc

In this case the two algorithms RA1 and RA2 coincide since there is only one outgoing arc.

We fix a node v^e with $n-1$ incoming arcs and 1 outgoing one and a Riemann initial datum $(\rho^0,\mu^0) = (\rho^{1,0},\mu^{1,0},\dots,\rho^{n-1,0},\mu^{n-1,0},\rho^{n,0},\mu^{n,0})$ as depicted in Figure 6.19. Let us denote with $(\hat{\rho},\hat{\mu}) = (\hat{\rho}^1,\hat{\mu}^1,\dots,\hat{\rho}^{n-1},\hat{\mu}^{n-1},\hat{\rho}^n,\hat{\mu}^n)$ the solution of the Riemann problem. In order to solve the dynamics we have to introduce the priority parameters $(q_1,q_2,\dots,q_{n-1})$ which determine a *level of priority* at the junction of incoming arcs.

Let us define

$$\Gamma_{inc} = \sum_{i=1}^{n-1} f_{\max}^i,$$

$$\Gamma_{out} = f_{\max}^n,$$

and $\Gamma = \min\{\Gamma_{inc},\Gamma_{out}\}$.

We analyze for simplicity the case in which $n=3$; in this case we need only one priority parameter $q \in]0,1[$. Think, for example, of a filling station for soda cans. Arc 3 fills the cans, whereas arcs 1 and 2 produce plastic and aluminium cans, respectively.

First, we compute $\hat{f}^e$, $e=1,2,3$, and then $\hat{\rho}^e$ and $\hat{\mu}^e$, $e=1,2,3$.

We have to distinguish two cases.

Case 1 $\Gamma = \Gamma_{inc}$;

Case 2 $\Gamma < \Gamma_{inc}$.

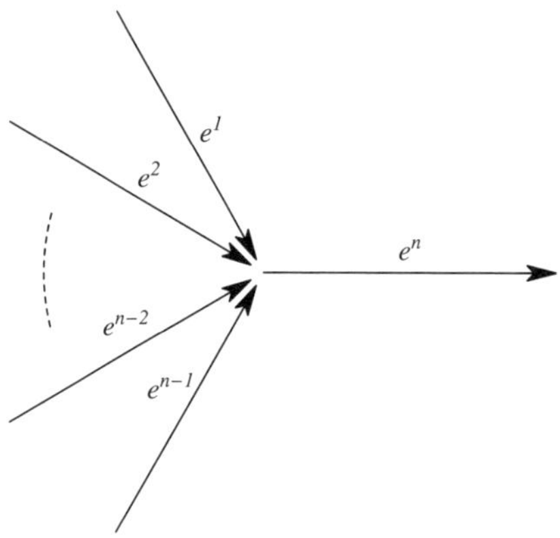

Figure 6.19. *One outgoing arc.*

In the first case we set $\hat{f}^i = f^i_{\max}$, $i = 1,2$. Let us analyze the second case in which we use the priority parameter q.

Not all objects can enter the junction, so let C be the number of objects that can go through. Then qC objects come from the first arc and $(1-q)C$ objects from the second. Consider the space $\left(f^1, f^2\right)$ and define the following lines:

$$r_q : f^2 = \frac{1-q}{q} f^1,$$

$$r_\Gamma : f^1 + f^2 = \Gamma.$$

Define P to be the point of intersection of the lines r_q and r_Γ. Recall that the final fluxes should belong to the region (see Figure 6.20)

$$\Omega = \left\{ \left(f^1, f^2\right) : 0 \le f^i \le f^i_{\max}, i = 1,2 \right\}.$$

We distinguish two cases:

(a) P belongs to Ω;

(b) P is outside Ω.

In the first case we set $(\hat{f}^1, \hat{f}^2) = P$, while in the second case we set $(\hat{f}^1, \hat{f}^2) = Q$, with $Q = proj_{\Omega \cap r_\Gamma}(P)$, where $proj$ is the usual projection on a convex set; see Figure 6.20.

Notice that $\hat{f}^3 = \Gamma$.

Remark 6.10.1. *The reasoning can be repeated also in the case of $n-1$ incoming arcs. In $\mathbb{R}^{n-1}$ the line r_q is given by $r_q = th_q$, $t \in \mathbb{R}$, with $h_q \in \Delta_{n-2}$, where*

$$\Delta_{n-2} = \left\{ (f^1, \dots, f^{n-1}) : f^i \ge 0, i = 1, \dots, n-1, \sum_{i=1}^{n-1} f^i = 1 \right\}$$

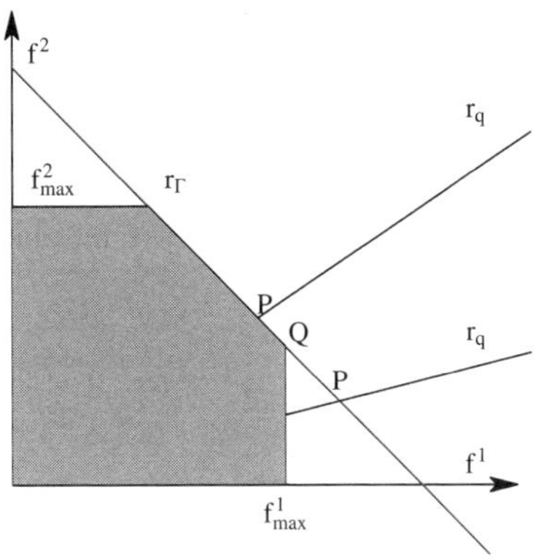

Figure 6.20. *P belongs to Ω, and P is outside Ω.*

is the $(n-2)$-dimensional simplex and

$$H_\Gamma = \left\{ (f^1, \dots, f^{n-1}) : \sum_{i=1}^{n-1} f^i = \Gamma \right\}$$

is a hyperplane. Since $h_q \in \Delta_{n-2}$, there exists a unique point $P = r_q \cap H_\Gamma$. If $P \in \Omega$, then we set $(\hat{f}^1, \dots, \hat{f}^{n-1}) = P$. If $P \notin \Omega$, then we set $(\hat{f}^1, \dots, \hat{f}^{n-1}) = Q = proj_{\Omega \cap H_\Gamma}(P)$, the projection over the subset $\Omega \cap H_\Gamma$. Observe that the projection is unique since $\Omega \cap H_\Gamma$ is a closed convex subset of H_Γ.

Let us compute $\hat{\rho}^e$ and $\hat{\mu}^e$, $e = 1,2,3$.
On the incoming arcs we have to distinguish two subcases.

Case 2.1 $\hat{f}^i = f^i_{\max}$. We set, according to rules SC2 and SC3,

$$\text{SC2}: \begin{array}{l} \hat{\rho}^i = \bar{\mu}^i, \\ \hat{\mu}^i = \bar{\mu}^i, \end{array} \qquad i = 1,2,$$

$$\text{SC3}: \begin{array}{l} \hat{\rho}^i = \bar{\mu}^i, \\ \hat{\mu}^i = \max\{\bar{\mu}^i, \mu^{i,0}\}, \end{array} \qquad i = 1,2.$$

Case 2.2 $\hat{f}^i < f^i_{\max}$. In this case there exists a unique $\hat{\mu}^i$ such that $\hat{\mu}^i + \varepsilon(\varphi(\hat{\mu}^i) - \hat{\mu}^i) = \hat{f}^i$. According to (6.45), we set $\hat{\rho}^i = \varphi(\hat{\mu}^i)$, $i = 1,2$.

Observe that in Case 2.1 $\hat{\rho}^i = \varphi(\hat{\mu}^i) = \bar{\mu}^i$, $i = 1,2$.
On the outgoing arc we have

$$\hat{\mu}^3 = \mu^{3,0},$$

while $\hat{\rho}^3$ is the unique value such that $f_\varepsilon(\mu^{3,0}, \hat{\rho}^3) = \hat{f}^3$.

One Incoming Arc

We fix a node v^e with 1 incoming arc and $m-1$ outgoing ones and a Riemann initial datum $(\rho^0,\mu^0)=(\rho^{1,0},\mu^{1,0},\rho^{2,0},\mu^{2,0},\dots,\rho^{m,0},\mu^{m,0})$ as depicted in Figure 6.21. Let us denote the solution of the Riemann problem by $(\hat{\rho},\hat{\mu})=(\hat{\rho}^1,\hat{\mu}^1,\hat{\rho}^2,\hat{\mu}^2,\dots,\hat{\rho}^m,\hat{\mu}^m)$. Since we have more than one outgoing arc, we need to define the distribution of goods from the incoming arc.

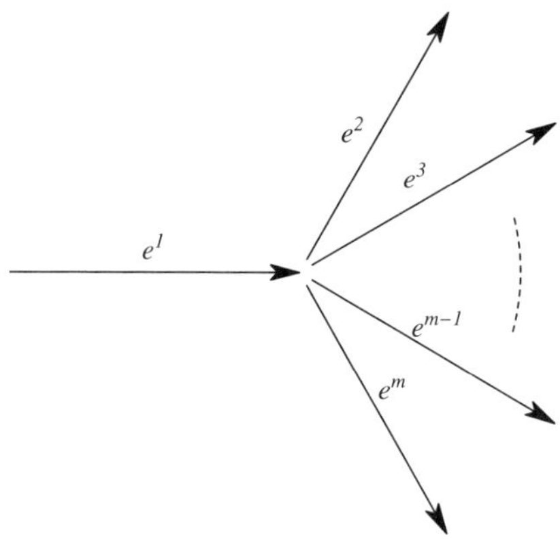

Figure 6.21. *One incoming arc.*

Introduce the flux distribution parameters $\alpha_j, j=2,\dots,m$, where

$$0<\alpha_j<1, \quad \sum_{j=2}^{m}\alpha_j=1.$$

The coefficient α_j denotes the percentage of objects addressed from the arc 1 to the arc j. The flux on the arc j is thus given by

$$f^j=\alpha_j f^1, \quad j=2,\dots,m,$$

where f^1 is the incoming flux on the arc 1.

Let us define

$$\Gamma_{inc}=f^1_{\max},$$

$$\Gamma_{out}=\sum_{j=2}^{m}f^j_{\max},$$

and $\Gamma=\min\{\Gamma_{inc},\Gamma_{out}\}$.

We have to determine $\hat{\mu}^e$ and $\hat{\rho}^e$, $e=1,\dots,m$, for both algorithms RA1 and RA2.

Riemann Solver According to RA1

Analyze the general case with m arcs. Consider, for example, the filling station for wine bottles of Section 6.1. The arcs e^3 and e^4 fill bottles with white and red wines, respectively, according to the bottle shapes. The dynamics at node v^2 is solved using the algorithm we are going to describe. Since $\hat{f}^j \leq f^j_{\max}$ it follows that

$$\hat{f}^1 \leq \frac{f^j_{\max}}{\alpha_j}, \quad j = 2, \ldots, m.$$

We set

$$\begin{aligned} \hat{f}^1 &= \min\{f^1_{\max}, \tfrac{f^j_{\max}}{\alpha_j}\}, \\ \hat{f}^j &= \alpha_j \hat{f}^1, \end{aligned} \qquad j = 2, \ldots, m.$$

On the incoming arc we have to distinguish two subcases.

Case 1 $\hat{f}^1 = f^1_{\max}$. According to rules SC2 and SC3, respectively, we set

$$\text{SC2}: \quad \begin{aligned} \hat{\rho}^1 &= \bar{\mu}^1, \\ \hat{\mu}^1 &= \bar{\mu}^1, \end{aligned}$$

$$\text{SC3}: \quad \begin{aligned} \hat{\rho}^1 &= \bar{\mu}^1, \\ \hat{\mu}^1 &= \max\{\bar{\mu}^1, \mu^{1,0}\}. \end{aligned}$$

Case 2 $\hat{f}^1 < f^1_{\max}$. In this case there exists a unique $\hat{\mu}^1$ such that $\hat{\mu}^1 + \varepsilon(\varphi(\hat{\mu}^1) - \hat{\mu}^1) = \hat{f}^1$. According to (6.45), we set $\hat{\rho}^1 = \varphi(\hat{\mu}^1)$.

On the outgoing arc we have

$$\hat{\mu}^j = \mu^{j,0}, \quad j = 2, 3,$$

while $\hat{\rho}^j$ is the unique value such that $f_\varepsilon(\mu^{j,0}, \hat{\rho}^j) = \hat{f}^j$, $j = 2, 3$.

Riemann Solver According to RA2

Let us analyze for simplicity the case in which $m = 3$; in this case we need only one distribution parameter $\alpha \in]0, 1[$. Think, for example, of the supply network of beach production described in Section 6.1. The dynamics at the node is solved according to algorithm RA2. Compute $\hat{f}^e$, $e = 1, 2, 3$.

We have to distinguish two cases.

Case 1 $\Gamma = \Gamma_{out}$;

Case 2 $\Gamma < \Gamma_{out}$.

In the first case we set $\hat{f}^j = f^j_{\max}$, $j = 2, 3$. Let us analyze the second case in which we use the priority parameter α.

Not all objects can enter the junction, so let C be the number of objects that can go through. Then αC objects come from the first arc and $(1-\alpha)C$ objects from the second. Consider the space (f^2, f^3) and define the following lines:

$$r_\alpha : f^3 = \frac{1-\alpha}{\alpha} f^2,$$

$$r_\Gamma : f^2 + f^3 = \Gamma.$$

Define P to be the point of intersection of the lines r_α and r_Γ. Recall that the final fluxes should belong to the region

$$\Omega = \left\{ \left(f^2, f^3\right) : 0 \le f^j \le f^j_{\max}, j = 2,3 \right\}.$$

We distinguish two cases:

(a) P belongs to Ω;

(b) P is outside Ω.

In the first case we set $(\hat{f}^2, \hat{f}^3) = P$, while in the second case we set $(\hat{f}^2, \hat{f}^3) = Q$, with $Q = proj_{\Omega \cap r_\Gamma}(P)$, where $proj$ is the usual projection on a convex set. Observe that $\hat{f}^1 = \Gamma$.

Again, we can extend the reasoning to the case of $m-1$ outgoing arcs as for the incoming arcs defining the hyperplane

$$H_\Gamma = \left\{ (f^2, \dots, f^m) : \sum_{j=2}^{m} f^j = \Gamma \right\}$$

and choosing a vector $h_\alpha \in \Delta_{m-2}$. Moreover, we compute $\hat{\rho}^e$ and $\hat{\mu}^e$ in the same way described for the Riemann Solver RA1.

Remark 6.10.2. *An alternative way of choosing the vector h_α is the following. We assume that a traffic distribution matrix A is assigned; then we compute $\hat{f}^1$ and choose $v_\alpha \in \Delta_{m-2}$ by*

$$h_\alpha = \Delta_{m-2} \cap \left\{ tA(\hat{f}^1) : t \in \mathbb{R} \right\}.$$

Remark 6.10.3. *The classical Kruzkov entropy inequalities at nodes* [10] *read*

$$\sum_{inc} sgn(\rho - e)(f(\rho) - f(e)) \ge \sum_{out} sgn(\rho - e)(f(\rho) - f(e)),$$

where the sums are, respectively, over incoming and outgoing arcs and e is arbitrary. The fluxes are always monotone with respect to ρ, while the precise values taken by fluxes and densities on the arcs may be different. Thus we cannot expect the inequality to hold in general.

Bullwhip Effect

The bullwhip effect is a well-known oscillation phenomenon in supply chain theory; see [24, 43, 44, 64, 65, 72]. Since the effect consists of oscillations moving backwards, we restrict ourselves to the most interesting case of nodes with $n-1$ incoming arcs and one outgoing arc.

To study the bullwhip effect, we compute the oscillations on incoming arcs produced by the interaction with the node of a wave from the outgoing one. Since the wave must have negative speed, it is a first family wave. To fix notation, let (ρ^-,μ^-) be an equilibrium configuration at the node and $((\rho^{n,-},\mu^{n,-}),(\tilde{\rho}^n,\tilde{\mu}^n))$ the wave coming to the node. In general, we denote with $-$ and $+$ the values before and after the interaction, while by Δ we indicate the jump in the values from the left to the right along waves travelling on arcs. The effect of the interaction of the wave is the production of $n-1$ waves on the incoming arcs.

The oscillation amplitude in the production rate before the interaction is given by

$$\Delta\mu^- = \tilde{\mu}^n - \mu^{n,-}.$$

The maximum flux on the outgoing arc as a function of μ is the following:

$$f^n_{\max}(\mu) = \mu\frac{1-\varepsilon}{1+\varepsilon} + \varepsilon\rho^n_{\max};$$

thus it is an increasing function. It follows that the oscillation of the flux after the interaction is

$$\Delta f^+ = \frac{1-\varepsilon}{1+\varepsilon}\Delta\mu^-.$$

Assume first that the incoming arcs are not active constraints. Then for both rules SC2 and SC3, we have $\rho^{i,-} \geq \mu^{i,-}, i = 1,\dots,n-1$. Then the first family curve passing through $(\rho^{i,-},\mu^{i,-})$, belonging to the region $\rho \geq \mu$, is given by

$$\rho = \rho^{i,-} + (\mu - \mu^{i,-})\left(-\frac{1-\varepsilon}{1+\varepsilon}\right),$$

from which, for small oscillations, we obtain

$$\Delta\rho^+ = -\frac{1-\varepsilon}{1+\varepsilon}\Delta\mu^+.$$

If the oscillation is not small, the same relation holds with an inequality sign. Observe that

$$\Delta f^+ = \Delta\mu^+(1-\varepsilon) + \varepsilon\Delta\rho^+ = \frac{1-\varepsilon}{1+\varepsilon}\Delta\mu^+,$$

from which

$$\Delta\mu^+ = \frac{1+\varepsilon}{1-\varepsilon}\Delta f^+,$$

and then

$$\Delta\mu^+ = \Delta\mu^-.$$

Assume now that the incoming arcs are active constraints. This means that $\mu^{i,-} = \rho^{i,-}$ for rule SC2 and $\mu^{i,-} \geq \rho^{i,-}$ for rule SC3. Along the curve of the first family belonging to the region $\rho \leq \mu$ we have $\Delta f = 0$; i.e., a dumping effect is possible. On the contrary, in the region $\rho \geq \mu$ we have

$$\Delta f = \frac{1-\varepsilon}{1+\varepsilon}\Delta\mu.$$

Consider first the case of rule SC2. In case the first family wave from the outgoing road increases the flux, then it is reflected as a second family wave. In the opposite case, we get the same estimates as above.

Consider now the case of rule SC3. In case the first family wave from the outgoing road increases the flux, then it is again reflected as a second family wave. In the opposite case, we get

$$\Delta\mu^{+} = \Delta\mu^{-} + (\mu^{i,-} - \rho^{i,-})$$

with an increase in the production rate oscillation.

Concluding we get the following proposition.

Proposition 6.24. *Rule SC3 may produce the bullwhip effect. On the contrary, rule SC2 conserves oscillations or produces a dumping effect, thus not permitting the bullwhip effect.*

6.11 Summary

In this chapter we have proposed a mixed continuum-discrete model; i.e., the supply chain is described by continuous arcs and discrete nodes, which means that the load dynamics is solved in a continuous way on the arcs, and at the nodes imposing conservation of goods density, but not of the processing rate. In fact, each arc is modelled by a system of two equations: a conservation law for the goods density, and an evolution equation for the processing rate. It is evident that this model and the model described in Chapter 5, with an ordinary differential equation for the queue, complete each other. In fact, the second model is more suitable when the presence of queue with buffer is fundamental to manage good production. The mixed continuum-discrete model, on the other hand, is useful when there is the possibility to reorganize the supply chain: particularly, the productive capacity can be readapted for some contingent necessity. Possible choices of solutions at nodes guaranteeing the conservation of fluxes are analyzed. In particular, Riemann Solvers are defined fixing the rules SC1, SC2, and SC3, and equilibria at nodes for all the rules are discussed. The numerical experiments described in Chapter 8 will show that SC1 appears to be very conservative (as expected), while SC2 and SC3 are more elastic, thus allowing richer dynamics. Then, the main difference between SC2 and SC3 is the following. SC2 tends to make adjustments of the processing rate more than SC3, even when it is not necessary for the purpose of flux maximization. Thus, when oscillating waves reach an arc, SC2 reacts by cutting such oscillations. In conclusion, SC3 is more appropriate to reproduce also the well-known bullwhip effect. For the Riemann Solver corresponding to rule SC1 we proved existence of solutions via accurate estimates of total variation of $f(\rho,\mu)$ (for BV initial data). As it happens for other macroscopic models on networks, a direct estimate on the total variation of ρ is not possible. In our case, it is interesting to notice that, also inside the arcs, a control of the total variation of the flux is obtained reasoning only on the special Riemann Solver chosen. Also, $T.V.(\rho)$ is of the order $(1/\varepsilon)T.V.(f)$; thus we have no uniform bound for arbitrary ε. More precisely, we have shown an example of unbounded increase of $T.V.(\rho)$ with respect to the initial datum; i.e., $T.V.(\rho(t))/T.V.(\rho(0))$ cannot be uniformly bounded with respect to ε.

The continuum-discrete model, regarding sequential supply chains, has been extended to supply networks with nodes of types $1 \times n$ and $m \times 1$. The Riemann problems are solved fixing two routing algorithms RA1 and RA2, already used for the analysis of packet flows in telecommunication networks. For both routing algorithms the flux of goods is maximized considering one of the two additional rules, SC2 and SC3. In the case of nodes with more incoming arcs and one outgoing arc, we have shown that SC3 reproduces the well-known bullwhip effect. The latter confirms that SC3 appears to be the more realistic modeling choice.

Chapter 7

Control and Optimization Problem for Networks

In the economic literature, the optimization of supply chains is mainly based on mixed-integer programming models; see [69, 74]. The new aspect is to introduce a connection between continuous network models and mixed-integer models. This relationship has already been investigated in the field of traffic flow and the detection of drinking water systems. For instance, in [32, 33], the starting point has been a continuous network model based on nonlinear partial differential equations (Lighthill/Whitham) for the evolution of traffic flow. At each intersection, the entries of the traffic distribution matrix are modeled as controls. The detailed model is then a coupled system of nonlinear partial differential equations with nonlinear constraints at the road intersections. The task is now to utilize the traffic network in such a way that all cars reach their final destination at the earliest possible time. To reformulate the nonlinearities into a mixed-integer framework, linearizations are used, and thus binary variables are generated.

From a mathematical point of view, there exist several approaches for solving partial differential equation constrained optimization problems: On the one hand, there is the treatment of the optimization problem in the partial differential equation context using adjoint equations deduced from the Lagrange principle. And, on the other hand, there is the possibility to derive a simplified model where the resulting discretized equations are interpreted as a mixed-integer problem (MIP) relating the partial differential equation scale to an MIP. We will see that the partial differential equation approach gives a guideline to develop new and dynamically more accurate models, whereas the usual MIP can be viewed as a simple and very coarse (two-point) approximation of the partial differential equation. From the viewpoint of the operations management literature advantages are the modeling in a strictly linear framework and the possibility to use fast and commercial solvers ensuring global extrema. Moreover, the continuous optimization approach opens a way to introduce nonlinearities in a straightforward and consistent way into these models and to treat nonlinear problems like chemical production lines by appropriate methods. A further ingredient is to introduce a relationship between continuous optimization models and mixed-integer programming models. For that purpose, we derive the continuous optimality system (necessary conditions for an extremum) and show that the mixed-integer formulation is also a valid discretization of the discretized continuous optimality system; i.e., both approaches "discretize-then-optimize" and "optimize-then-discretize" lead to the same continuous optimal control if the discretization width tends to zero. In fact, the advantage of this MIP approach to partial differential equation models is twofold. First, in many cases MIPs can be solved even for large-scale instances in reasonable time by state-of-the-art numerical solvers. Second, the solutions come with a quality warranty; that is, either optimality is proved or an estimation of the optimality gap for the best-known solution is returned. Linear mixed-integer programming can be applied in many ways to continuous partial differential equation models. More generally, we show a connection between the partial differential equation approach and the MIP approach, which yields (dependent on the discretization) the same model but differs by the solution method (combinatorial methods versus gradient computations for adjoint systems).

This chapter is organized as follows. We will introduce the continuous optimal control problem for supply chains that is to be investigated; see Section 7.2. In Section 7.3, we derive the optimality system for both the continuous and the discrete supply chain model. Afterwards, in Section 7.4, we discuss the discretization leading to the mixed-integer model that is, however, transformed into a linear program under certain assumptions in Section 7.5.

7.1 Motivation

The formulation of optimal control problems for supply chain management is an immediate consequence of successful simulations with partial differential equations. Having results on the behavior of the solutions to the models at hand, one should ask for adjusting parameters in the models in an "optimal" way. In the context of supply chain management this may include the question of optimal processing parameters, minimizing buffering queues to reduce costs or to ensure fully loaded production lines. More precisely, depending on the actual application, several questions are of importance: What is the maximum value the queues attain? Can we control the distribution such that we achieve a maximal outflow? What is the maximal load of the factory? How long does it take to process a certain good? All these items indirectly imply the problem of optimal routing of goods through networks. For instance, the supply chain as illustrated in Figure 7.1 permits four alternatives (marked as gray boxes) to influence distributing the goods.

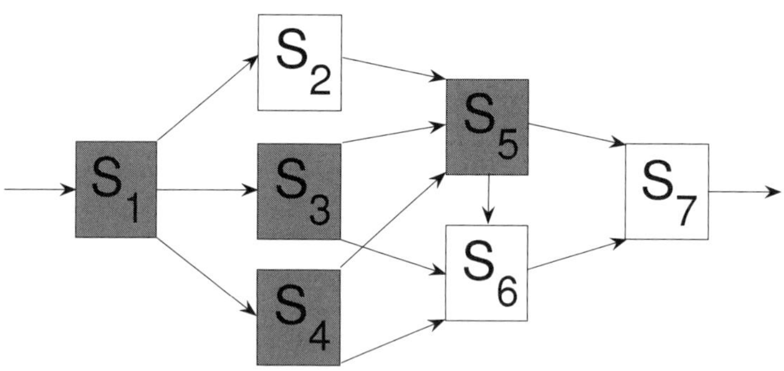

Figure 7.1. *Example of a production network.*

The question we ask is how to control the flow through the network so that a maximum number of goods can be produced and storage costs are minimized. In a simulation only the product flow propagates according to user-defined distribution rates. The simplest choice is equally distributed rates; i.e., the arriving goods are consistently fed in parallel machines. However, this may result in an inefficient load of queues and does not guarantee the greatest possible output; cf. Figure 7.2. Unlike this way, optimization procedures seek to operate the production network at minimum queue costs and high output. In particular, for time-varying inflows of raw materials and highly connected suppliers this leads to a challenging mathematical problem.

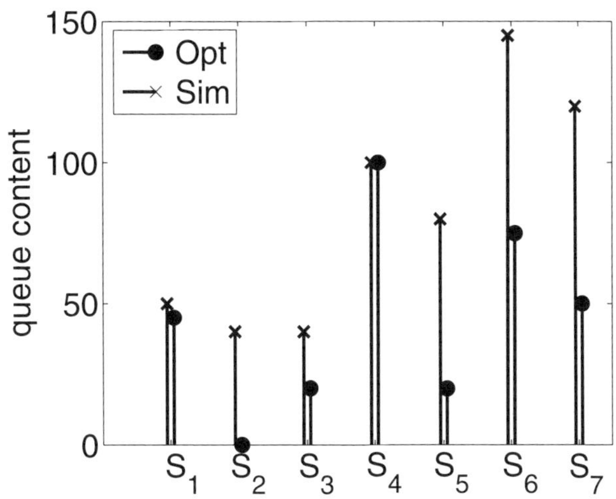

Figure 7.2. *Simulation versus optimization results.*

We are concerned with the optimization of production systems with special focus on synergies and similarities between different mathematical formulations. Our principal object is to provide a framework that is suitable for concepts of well-known optimization procedures.

7.2 Basics

The starting point for our investigations is the continuous network model introduced in Section 5.2. Below we rely on the already proposed model and describe its corresponding optimal control problem. In particular, we use a simplification of the queue outflow in order to prevent the discontinuous dependence on the queue length; cf. (5.6), (5.7), and (5.13). The advantage of this reformulation is twofold: The resulting MIP should contain as few as possible binary variables to reduce complexity. Otherwise, to apply adjoint calculus it is mandatory to require the constraints to be differentiable. For a detailed interpretation of the adapted outflow (5.7) we refer the reader to Section 5.3, where the network model is reformulated in a clever way. Note that this queue outflow is nothing but a fallout of discretizing (5.1) and considering the limit $t \to \infty$ of (5.13).

First, let us give a brief idea of the detailed discussion: We assume that a general supply chain is represented as a network of directed arcs. On each of its arcs the dynamics of the density evolution of the processed parts is given by a linear advection equation. At each vertex of the network we can influence the distribution of the mass flux of the parts by a control. Also, different arc dynamics are coupled at a vertex by an ordinary differential equation. Finally, we measure some entities on the whole network and obtain the optimal control problem by asking for a control such that the dynamics on arcs and vertices yield a maximal entity.

Remember that each vertex models a merge or disperse of the supply chain. In the case of a dispersing intersection we introduce time-dependent distribution rates $A^{v,e}(t)$, $v \in \mathcal{V}_d$, with $\mathcal{V}_d$ being the set of all dispersing junctions. The distribution rate $A^{v,e}(t)$ describes the percentage of the flux coming from prior processors going to the queue of processor e. The functions $A^{v,e}$ are required to satisfy $0 \leq A^{v,e}(t) \leq 1$ and $\sum_{e \in \delta_v^+} A^{v,e}(t) = 1$ for all times t. In the following, the entries of the distribution vectors $A^{v,e}(t), v \in \mathcal{V}_d$, are the controls for the supply chain process described above. We determine the optimal controls $A^{v,e}(t)$ as solutions to an optimal control problem. The optimization problems under consideration consist of partial as well as ordinary differential and algebraic equations as constraints. Each evaluation of the cost functional requires in particular the solution of the partial differential equation on the whole network. Usually, a numerical solution of the problem will be obtained by a suitable discretization of the partial and ordinary differential equations and the application of an appropriate nonlinear optimization method such as an adjoint calculus; see Section 7.3. Depending on the discretization in space and time, a hierarchy of different models ranging from a very coarse two-point discretization for each arc, which is fast but not an accurate solution of the partial differential equation, to a very fine discretization of the dynamics on each arc being accurate but very expensive is obtained. The cost functional

$$\min_{A^{v,e}(t), v \in \mathcal{V}_d} \sum_{e \in \mathcal{A}} \int_0^T \int_{a^e}^{b^e} \mathcal{F}(\rho^e(t,x), q^e(t))\, dx\, dt \tag{7.1}$$

will be chosen with respect to the real-world applications. One possibility consists in using

$$\mathcal{F}(\rho^e, q^e) = \rho^e(t,x) + \frac{1}{L^e} q^e(t), \tag{7.2}$$

which measures the entity of goods per time unit, but we point out the fact that the following discussion proceeds analogously for different, but separable (enables the evaluation on each arc), cost functionals. The latter (7.2) can be interpreted as minimizing the load of the network.

Remark 7.2.1. *There is also the possibility to choose the processing velocities V^e as control parameters. The number of controls is then dependent on the network size, more precisely on the number of arcs, since the velocities are constant in space and time. Note that the optimization problem becomes nonlinear with the choice of these controls.*

7.3 A General Adjoint Calculus for Transport Processes on Networks

In this section we present a calculus to efficiently solve optimization problems arising in supply chain networks. It is well known that for the given optimization problems there exists a necessary first-order optimality system or Karush–Kuhn–Tucker (KKT) system (under suitable assumptions on the appearing functions). The presented calculus is hence also based on these equations. However, due to the special structure of the network problem and the applicable controls we can exploit structural information in the first-order optimality system. This allows later for a fast and efficient solution of the necessary optimality conditions.

As in the previous section we study the following optimization problem on a network graph $(\mathcal{A},\mathcal{V})$ for given initial data ρ_I^e and q_I^e on a finite time horizon T:

$$\min_{A^{v,e}(t),v\in\mathcal{V}_d} \sum_{e\in\mathcal{A}} \int_0^T \int_{a^e}^{b^e} \mathcal{F}(\rho^e(t,x),q^e(t))\,dx\,dt \tag{7.3a}$$

$$\text{subject to } e\in\mathcal{A},\quad v\in\mathcal{V},\quad t\in(0,T),\quad x\in[a^e,b^e], \tag{7.3b}$$

$$\partial_t\rho^e+V^e\partial_x\rho^e=0,\quad \rho^e(0,x)=\rho_I^e(x),\quad V^e\rho^e(t,a^e)=f^e(q^e(t)), \tag{7.3c}$$

$$\partial_t q^e(t)=A^{v,e}(t)\left(\sum_{\tilde{e}\in\delta_v^-} V^{\tilde{e}}\rho^{\tilde{e}}(t,x_v^{\tilde{e}})\right)-f^e(q^e(t)),\quad q^e(0)=q_I^e, \tag{7.3d}$$

$$f^e(q^e(t))=\min\left\{\mu^e;\frac{q^e(t)}{\epsilon}\right\}. \tag{7.3e}$$

Classically, this problem would be viewed as a constraint optimization problem where the constraints are ordinary and partial differential equations. However, this problem also allows for a different interpretation as an optimal control problem in the following sense: The quantities of interest are the distribution rates $A^{v,e}(t)$ at each node v of the network. Once we know these quantities we can evaluate the model and obtain ρ^e and q^e as well-defined solutions to the governing supply chain equations (7.3b)–(7.3e). The idea is to exploit this structural information later on for optimization purposes. For the subsequent analysis and formal derivation we introduce the following notation. We assume that all arcs are labeled $e=1,\ldots,M$. All controls $(A^{v,e})_{v,e}$ are collected in a sufficiently large vector u; i.e., we have one control for each arc e and set $u=(A^{v,e})_e$. The flux function f^e will be denoted by $f^e(q)$, and it is assumed that it is evaluated only at t and q^e. Without loss of generality, we assume that $a^e=0$ and $b^e=1$. We denote the conservation laws for the $\rho=(\rho^e)_e$ in (7.3c) by

$$C_\rho(\rho,q,u)=0$$

and the ordinary differential equations $q=(q^e)_e$ in (7.3d) for the buffers by

$$C_q(\rho,q,u)=0.$$

The functional is denoted by G, and this may depend on ρ,q and possibly also on the controls u. Thus, we have to solve the constrained optimization problem

$$G(\rho,q,u)\to\min,\quad C_\rho(\rho,q,u)=0,\quad C_q(\rho,q,u)=0 \tag{7.4}$$

for the given functional $G(\rho,q,u)$.We also might replace (7.4) by minimizing the unconstrained functional (Lagrangian)

$$G(\rho,q,u)+\langle\psi,C_\rho(\rho,q,u)\rangle+\langle\phi,C_q(\rho,q,u)\rangle\to\min \tag{7.5}$$

over all possible functions ρ,q,u,ψ,ϕ, where $\psi^e(t,x)$ is interpreted as the Lagrange multiplier for ρ^e in (7.3c) and $\phi^e(t)$ is the Lagrange multiplier for q^e in (7.3d). The functional stated in (7.5) is the Lagrangian for the constrained problem, and the definition of the dual product $\langle\cdot,\cdot\rangle$ is made precise below.

In order to use the Lagrange multiplier approach, we first have to reformulate the constraints (7.3c)–(7.3d) in a weak form. We choose a fixed time horizon T, label the arcs as usual by $e=1,\dots,M$, and define

$$C_\rho(\rho,q,u)=0 \iff \langle\psi,C_\rho(\rho,q,u)\rangle=0 \quad \forall\psi,$$

where the scalar product $\langle *,*\rangle_\rho$ is defined by

$$\langle\psi,C_\rho\rangle_\rho=\sum_{e\in\mathcal{A}}\Bigg\{\int_0^1 \psi^e(T,x)\rho^e(T,x)-\psi^e(0,x)\rho_I^e(x)\,dx \tag{7.6}$$

$$+\int_0^T V^e\psi^e(t,1)\rho^e(t,1)-\psi^e(t,0)f^e(q^e)\,dt-\int_0^T\int_0^1 \partial_t\psi^e\rho^e+V^e\partial_x\psi^e\rho^e dxdt\Bigg\}.$$

Similarly, we define the constraints $C_q(\rho,q,u)=0$ in (7.3d) weakly by

$$C_q(\rho,q,u)=0 \iff \langle\phi,C_q(\rho,q,u)\rangle_q=0 \quad \forall\phi,$$

where the scalar product $\langle *,*\rangle_q$ is defined by

$$\langle\phi,C_q(\rho,q,u)\rangle_q \tag{7.7}$$

$$=\sum_{e\in\mathcal{A}}\Bigg\{\phi^e(T)q^e(T)-\phi^e(0)q_I^e+\int_0^T -\partial_t\phi^e q^e+\phi^e[f^e(q)-u^e\sum_{\tilde e\in\delta_v^-}V^{\tilde e}\rho^{\tilde e}(t,1)]\,dt\Bigg\}.$$

Thus, we have M test functions, or Lagrange multipliers, $\psi^e(t,x)$ for the constraints $C_\rho$in (7.3c) and M Lagrange multipliers $\phi^e(t)$ for the constraints C_q in (7.3d). Optimizing the unconstrained functional in (7.5) now corresponds to setting all the Fréchet derivatives with respect to ρ,q,u,ψ,ϕ equal to zero.

This gives in weak form for variations $\delta\rho,\delta q$, etc., the following set of equations:

$$D_\rho G(\rho,q,u)\delta\rho+\langle\psi,D_\rho C_\rho(\rho,q,u)\delta\rho\rangle_\rho+\langle\phi,D_\rho C_q(\rho,q,u)\delta\rho\rangle_q=0 \quad \forall\delta\rho, \tag{7.8a}$$

$$D_q G(\rho,q,u)\delta q+\langle\psi,D_q C_\rho(\rho,q,u)\delta q\rangle_\rho+\langle\phi,D_q C_q(\rho,q,u)\delta q\rangle_q=0 \quad \forall\delta q, \tag{7.8b}$$

$$D_u G(\rho,q,u)\delta u+\langle\psi,D_u C_\rho(\rho,q,u)\delta u\rangle_\rho+\langle\phi,D_u C_\rho(\rho,q,u)\delta u\rangle_\rho=0 \quad \forall\delta u, \tag{7.8c}$$

$$C_\rho(\rho,q,u)=0, \quad C_q(\rho,q,u)=0. \tag{7.8d}$$

7.3.1 Necessary First-Order Conditions (Continuous Case)

Equations (7.8), determining the optimizer of the Lagrangian, the unconstrained functional in (7.5), can be rewritten in their strong form again, leading to the solution of a set of differential algebraic equations for the Lagrange multipliers ψ^e,ϕ^e and the controls u. Assuming sufficient regularity of the arising operators we can compute now $D_\rho G, D_q G, D_u G$

as linear operators into $\mathbb{R}$. For our particular choice of these operators and cost functional all derivatives can be expressed in terms of integral kernels. Denoting by $\delta\rho, \delta q$, and δu the variations with respect to states ρ, q and controls u and denoting the integral kernels by $K_\rho^e(x,t)$, $K_q^e(t)$, and $K_u^e(t)$, we obtain

$$D_\rho G(\rho,q,u)\delta\rho = \sum_{e\in\mathcal{A}} \int_0^T \int_0^1 K_\rho^e(\rho,q,u)\delta\rho^e \, dxdt, \tag{7.9a}$$

$$D_q G(\rho,q,u)\delta q = \sum_{e\in\mathcal{A}} \int_0^T K_q^e(\rho,q,u)\delta q^e \, dt, \tag{7.9b}$$

$$D_u G(\rho,q,u)\delta u = \sum_{e\in\mathcal{A}} \int_0^T K_u^e(\rho,q,u)\delta u^e \, dt. \tag{7.9c}$$

In the following we compute the strong form to (7.9). Starting with the equation for ψ we differentiate the unconstrained functional (7.5) with respect to ρ, using (7.6)–(7.7), and obtain

$$\sum_{e\in\mathcal{A}} \Bigg\{ \int_0^T \int_0^1 K_\rho^e(\rho,q,u)\delta\rho^e dxdt + \int_0^1 \psi^e \delta\rho^e(T,x)\, dx + \int_0^T V^e\psi^e(t,1)\delta\rho^e(t,1)\, dt$$

$$- \int_0^T \int_0^1 [\partial_t \psi^e + V^e \partial_x \psi^e]\delta\rho^e dxdt - \int_0^T \phi^e u^e \sum_{\tilde{e}\in\delta_v^-} V^{\tilde{e}}\delta\rho^{\tilde{e}}(t,1)\, dt \Bigg\} = 0 \quad \forall \delta\rho.$$

This is clearly the weak formulation of the terminal BV problem

$$\partial_t \psi^e + V^e \partial_x \psi^e = K_\rho^e(\rho,q,u), \tag{7.10a}$$

$$\psi^e(x,T) = 0, \quad \psi^e V^e(t,1) = \sum_{\tilde{e}\in\delta_v^+} \phi^{\tilde{e}}(t) u^{\tilde{e}} V^{\tilde{e}}. \tag{7.10b}$$

Similarly, we obtain an equation for ϕ by computing the strong form of (7.8b) and differentiating the unconstrained functional (7.5) with respect to q using (7.6)–(7.7). We obtain

$$\sum_{e\in\mathcal{A}} \Bigg\{ \int_0^1 \int_0^T K_q^e(\rho,q,u)\delta q^e \, dxdt - \int_0^T \psi^e(t,0)\partial_{q^e} f^e(q)\delta q^e \, dt + \phi^e(T)\delta q^e(T)$$

$$+ \int_0^T [-\partial_t \phi^e \delta q^e(t) + \phi^e \partial_{q_e} f^e(q)]\delta q^e \, dt \Bigg\} \quad \forall \delta q.$$

This is clearly the weak formulation of the terminal value problem

$$\partial_t \phi^e = [\phi^e - \psi^e(t,0)]\partial_{q^e} f^e(q^e) + \int_0^1 K_q^e(\rho,q,u)dx, \quad \phi^e(T) = 0. \tag{7.11}$$

Finally, we obtain the equation for u as

$$K_u^e(q,u,t) - \sum_{\tilde{e}\in\delta_v^-} \phi^{\tilde{e}} V^{\tilde{e}} \rho^{\tilde{e}}(t,1) = 0. \tag{7.12}$$

7.3.2 Numerical Solution of First-Order Optimality System (Continuous Case)

Solving the Lagrangian equations (7.8), using the form (7.10), (7.11), (7.12) of the individual equations is a quite formidable computational task, since we have doubled the number of simulation variables and have to deal with a coupled forward-in-time and backward-in-time problem. Moreover, the problem will in general have multiple solutions, for the simple reason that there will be regions in the space-time domain where the controls do not affect the solution. This implies that there might be in general many optimal controls. These problems can be avoided by computing an optimal control via, for example, the steepest descent method. As shown below, this method can also be seen as (7.8), a block Gauss–Seidel approach. We discuss the basic idea of deriving the steepest descent approach in this section and state the corresponding equations. The algorithm is summarized at the end. The following presentation is exemplified on the example of supply chains but can be and has been applied similarly to other problems such as traffic flow [46, 48], general partial differential equation constrained optimization problems [51, 73], and radiative transfer problems [49, 52].

We start the discussion noting that (7.8a)–(7.8d) are actually easily solved for a given control u since solving (7.8d) corresponds to just a forward simulation of the network, i.e., solving (7.3c)–(7.3d) for a given u. Once ρ and q are obtained this way, solving (7.8a), (7.8b), or (7.10), (7.11), respectively, corresponds to solving a problem of essentially the same structure and complexity backwards, i.e., to solving the dual network problem. This leaves only (7.8c), or (7.12), respectively, to deal with. We use the following idea. We formally denote the solution operator of the forward network problem (7.3c)–(7.3d) for a given u by $(\rho,q) = F(u)$. Solving the constrained optimization problem now formally corresponds to minimizing the unconstrained functional

$$g(u) = G(F(u),u).$$

To solve this problem by a steepest descent method, we would have to compute the gradient direction of g for a given iterate u; i.e., we would have to compute $D_u g$ given by

$$D_u g \delta u = D_{\rho q} G(F,u) D_u F(u) \delta u + D_u G(F,u) \delta u.$$

Note that g and G are functionals into the real numbers, and so their derivatives can be identified with functions. Hence,

$$\sum_{e\in\mathcal{A}} \int_0^T D_{u^e} g(t) \delta u^e(t)\, dt = \sum_{e\in\mathcal{A}} \int_0^T [D_{\rho q} G(F,u) D_u F(u) + D_u G(F,u)]_e(t) \delta u^e(t)\, dt \;\forall \delta u^e,$$

$$D_{u^e} g(t) = [D_{\rho q} G(F,u) D_u F(u) + D_u G(F,u)]_e(t).$$

Certainly, $F(u)$ and its derivative $D_u F$ are not explicitly available. Knowledge on F and DF would require having an explicit representation of the state (ρ,q) in terms of the control u. This is typically not the case whenever the constraints are given by partial or even ordinary differential equations.

However, it can be shown that *for a given control u the term* $[D_{\rho q} G(F,u) D_u F(u) + D_u G(F,u)]_e(t)$ *is the residual in* (7.8c) *when solving* (7.8a)–(7.8d). This is also true in more general situations; see, e.g., [73].

Remark 7.3.1. *This relation can be easily understood when considering the following similar but finite-dimensional example. Assume we want to solve the problem*

$$\min_{y,u} f(u,y) \quad \textit{subject to} \quad \hat{E}Ay = u \tag{7.13}$$

for $y = (\rho, q)$ being the state and u being the control. Both are assumed to be vectors in $\mathbb{R}^n$. A is a matrix in $\mathbb{R}^{n\times n}$ and f a differentiable possibly nonlinear function.

If we assume that $A \in \mathbb{R}^{n\times n}$ is invertible (this corresponds to assuming that for any control u there exists a unique solution (ρ, q) to the supply chain problem), then we can also study the reduced (unconstrained) problem

$$\min_{u} \tilde{f}(u). \tag{7.14}$$

Herein, $\tilde{f}(u) = f(u, y(u))$ and $y(u) = A^{-1}u$. The minimization problem is unconstrained, and therefore the necessary condition is

$$\nabla_u \tilde{f}(u) = 0. \tag{7.15}$$

Evaluating the gradient gives

$$D_u \tilde{f}(u) = D_u f(u,y) + D_y f(u,y) D_u y \textit{ or } \nabla_u \tilde{f}(u) = \nabla_u f + A^{-T} \nabla_y f(u,y). \tag{7.16}$$

Hence, we would need to compute $D_u y$ and A^{-T} by differentiating $y = A^{-1}u$ (this is typically impossible for problems involving partial differential equations as seen above). Assuming that the vector λ solves

$$\textit{adjoint equation } \nabla_y f(u,y) - A^T \lambda = 0, \tag{7.17}$$

and y solves

$$\textit{state equation } Ay = u, \tag{7.18}$$

we obtain

$$\nabla_u \tilde{f}(u) = \nabla_u f(u,y) + \lambda. \tag{7.19}$$

This is an equation which can be evaluated easily and yields a gradient for the reduced cost functional. Hence, in order to obtain a descent direction for a given control u we need to compute the state equation for a fixed control u and the adjoint equation for a fixed state y and control u and to evaluate $\nabla_u f(u,y) + \lambda$. These computations can be extended to the case A nonlinear and A given as an operator representing a partial differential equation. The difference is that the equation for λ will be a partial differential equation, too. In fact, the state and adjoint equations are the variations of the Lagrangian with respect to the multiplier and the state, respectively. These variations have been computed previously.

According to the calculations done in the previous remark, we propose the following steepest descent algorithm for solving problem (7.3).

- *Initialization*

  For a given control $u(t)$

- *State equation*

  Solve the network problem (7.8d); i.e., solve (7.3c)–(7.3d):

$$\text{(a) } \partial_t \rho^e + V^e \partial_x \rho^e = 0, \quad V^e \rho^e(t,0) = f^e(q^e), \quad \rho^e(0,x) = \rho_I^e(x),$$

$$\text{(b) } \partial_t q^e + f^e(q^e) - \sum_{e \in \mathcal{A}} (u^e) V^e \rho^e(t,1) = 0, \quad q^e(0) = q_I^e, \quad e = 1, \ldots, M.$$

- *Adjoint equation*

  Now, given ρ and q solve the adjoint network problem (7.8a)–(7.8b); i.e., solve (7.10) and (7.11) backward in time:

$$\text{(a) } \partial_t \psi^e + V^e \partial_x \psi^e = K_\rho^e(\rho, q, u),$$

$$\psi^e(T,x) = 0, \quad \psi^e V^e(t,1) = \sum_{\tilde{e} \in \delta_v^+} \phi^{\tilde{e}}(t)(u^{\tilde{e}}) V^{\tilde{e}}, \quad e = 1, \ldots, M,$$

$$\text{(b) } \partial_t \phi^e = [\phi^e - \psi^e(t,0)] \partial_{q^e} f^e(q^e) + \int_0^1 dx \;\; K_q^e(\rho, q, u),$$

$$\phi^e(T) = 0, \quad e = 1, \ldots, M.$$

- *Gradient equation*

  Compute the gradient direction $D_{u_e} g$ from the residual in (7.8c), according to (7.12):

$$D_{u^e} g(u) = K_{u^e}^e(q,u,t) + \sum_{e \in \mathcal{A}} [\phi^e - \psi^e(t,0)] \partial_{u^e} f^e(q^e) - \sum_{e \in \mathcal{A}} \sum_{\tilde{e} \in \delta_v^+} \phi^e u^{\tilde{e}} V^{\tilde{e}} \rho^{\tilde{e}}(t,1).$$

- *Repeat ...*

  Advance u in the direction of $-D_u g$ using a gradient-based optimization method [56]. The simplest possible method uses a fixed step size $\kappa > 0$ and updates the control by

$$u^{new} = u^{old} - \kappa D_u g.$$

  A more sophisticated approach chooses κ as large as possible, provided that there is still a sufficient decrease in the cost functional. Other methods are adapted from quasi-Newton methods and use information on an approximate Hessian. We refer the reader to [56] for an overview and more details. After advancing u we start over with the new control u^{new} and the first step of the algorithm.

Remark 7.3.2. *Note that gradient information could also be obtained by numerical means. One possibility would be to use, e.g., finite differences on the control. However, if we assume that u has length #u, and we use central differencing, then we need* 2#u *solutions to the full supply chain model* (7.3c)–(7.3d) *in order to compute a single gradient. If we assume that every solution of the supply chain model needs a CPU time c_1 and assuming that the adjoint equations need the same time, we can compare the computational time for calculating a gradient. In the case of finite differences it is about* 2#uc_1, *and with the previous algorithm it is* 2#u. *Since the gradient has to be computed in every descent step, there is a clear advantage in using the adjoint equations over finite differencing.*

7.3.3 Necessary First-Order Optimality Conditions (Discrete Case)

The previous derivation has been given independently of any numerical discretization. For numerical computations it is necessary to discuss the discrete case. Hence, the main purpose of this section is to show that the derived concept is also valid in the discrete case and can be applied in a similar fashion. Second, we discuss properties of the derived discretization. One possibility is to discretize the given first-order optimality conditions derived in the previous sections and solve them numerically. This procedure is known as the "optimize-then-discretize" approach. However, it is necessary to also analyze the following procedure: Discretize the supply chain model (7.3c)–(7.3d) and then optimize the discrete set of equations ("discretize-then-optimize"). In this section we will proceed using the second approach and show that, in fact, this yields (up to the order of the discretization error) exactly the same equations as the "optimize-then-discretize" approach.

We start by deriving the discrete optimality system. As in the previous section we consider the supply chain model as given by

$$\min_{A^{v,e}(t), v\in\mathcal{V}_d} \sum_{e\in\mathcal{A}} \int_0^T \int_{a^e}^{b^e} \mathcal{F}(\rho^e(t,x), q^e(t))\, dx\, dt \tag{7.20a}$$

$$\text{subject to } e\in\mathcal{A}, \quad v\in\mathcal{V}, \quad t\in(0,T), \quad x\in[a^e,b^e], \tag{7.20b}$$

$$\partial_t\rho^e + V^e\partial_x\rho^e = 0, \quad \rho^e(0,x) = \rho_I^e(x), \quad V^e\rho^e(t,a^e) = f^e(q^e(t)), \tag{7.20c}$$

$$\partial_t q^e(t) = A^{v,e}(t)\left(\sum_{\tilde{e}\in\delta_v^-} V^{\tilde{e}}\rho^{\tilde{e}}(t,x_v^{\tilde{e}})\right) - f^e(q^e(t)), \quad q^e(0) = q_I^e, \tag{7.20d}$$

$$f^e(q^e(t)) = \min\left\{\mu^e; \frac{q^e(t)}{\epsilon}\right\}. \tag{7.20e}$$

A coarse grid discretization in space of (7.20c) is obtained by just a two-point upwind discretization, and (7.20d) is discretized using the explicit Euler method. Each arc has length $L^e := b^e - a^e$, and we introduce a step size Δt such that step-size restrictions are met. The time steps t_j are numbered by $j = 0,\ldots,N_T$. In order to simplify the notation we use the following abbreviations for all e, j:

$$\rho_j^{e,b} := \rho^e(t_j, b^e), \quad \rho_j^{e,a} := \rho^e(t_j, a^e), \quad q_j^e := q^e(t_j), \quad A_j^{v,e} := A^{v,e}(t_j), \tag{7.21}$$

$$h_j^e := A^{v,e}(t_j)\left(\sum_{\tilde{e}\in\delta_v^-} V^{\tilde{e}}\rho^{\tilde{e}}(t_j, x_v^{\tilde{e}})\right). \tag{7.22}$$

Furthermore, we note that in the following calculations we need to differentiate the flux function f^e with respect to q^e. Of course, this is only possible in a strict way if we replace f^e by a smooth approximation of the minimum. However, at least formally, we can compute

$$(f^e)'(q^e) = \frac{1}{\epsilon} H\left(\mu^e - \frac{q^e}{\epsilon}\right),$$

where $H(\cdot)$ is the Heaviside function. Due to the boundary condition $V^e\rho^e(t,a^e) = f^e(q^e(t))$ we replace the discrete variable $\rho_j^{e,a}$ by $f^e(q_j^e)$, and therefore $\rho_j^{e,a}$ does not appear explicitly in the discrete optimal control problem below. For the initial data we have

$$\rho_0^{e,b} = \rho_0^{e,a} = q_0^e = 0 \quad \forall e. \tag{7.23}$$

Finally, the discretization of problem (7.3) reads, for $j \geq 1$, $e \in \mathcal{A}$, $v \in \mathcal{V}$,

$$\min_{A^{v,e}(t),\, v\in\mathcal{V}_d} \sum_{e\in\mathcal{A}} \sum_{j=1}^{N_T-1} \Delta t L^e \mathcal{F}(\rho_j^{e,a}, q_j^e) \tag{7.24a}$$

subject to

$$\rho_{j+1}^{e,b} = \rho_j^{e,b} + \frac{\Delta t}{L^e}(f^e(q_j^e) - V^e \rho_j^{e,b}), \tag{7.24b}$$

$$q_{j+1}^e = q_j^e + \Delta t(h_j^e - f^e(q_j^e)). \tag{7.24c}$$

Remark 7.3.3. *For deriving the discrete optimality system we state the precise definition of h^e in the case of the following intersections; see Figure* 7.3.

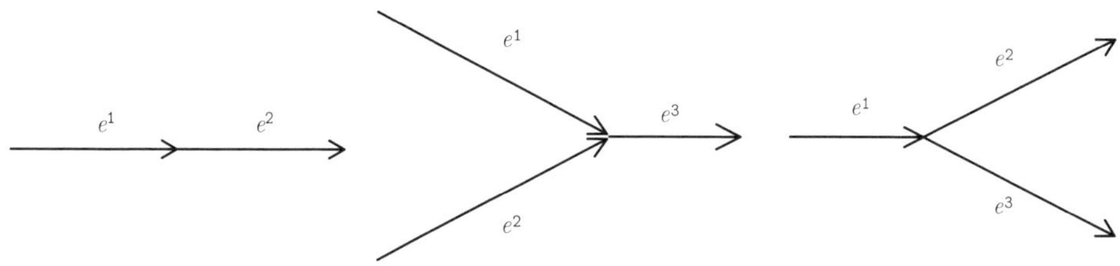

Figure 7.3. *Sample intersections labeled as types A, B, and C (from left to right).*

In case A, h^2 is independent of $A^{v,e}$ and we have $h^2(\vec{\rho}^e, A^{v,e}) = V^1\rho^1(b,t)$. Similarly, in case B we obtain $h^3(\vec{\rho}^e, A^{v,e}) = V^1\rho^1(b,t) + V^2\rho^2(b,t)$. Finally, as already stated, we have, in the controlled case C, $h^2(\vec{\rho}^e, A^{v,e}) = A^{v,2}(t)V^1\rho^1(b,t)$ and $h^3(\vec{\rho}^e, A^{v,e}) = (1 - A^{v,2}(t))V^1\rho^1(b,t)$.

Now, it is straightforward to derive the discrete optimality system for (7.24). We denote the Lagrange multipliers for the discretized partial differential equation by λ_j^e and for the discretized ordinary differential equation by p_j^e. The discrete Lagrangian is given by

$$L(\vec{\rho}_j^e, \vec{q}_j^e, A_j^{v,e}, \vec{\lambda}_j^e, \vec{p}_j^e) = \sum_{e\in\mathcal{A}} \sum_{j=1}^{N_T-1} \Delta t \left(L^e \mathcal{F}(\rho_j^{e,a}, q_j^e) \right) \tag{7.25a}$$

$$- \sum_{e\in\mathcal{A}} \sum_{j=1}^{N_T} \Delta t L^e \lambda_j^e \left(\frac{\rho_{j+1}^{e,b} - \rho_j^{e,b}}{\Delta t} - \frac{\psi(q_j^e) - V^e \rho_j^{e,b}}{L^e} \right) \tag{7.25b}$$

$$- \sum_{e\in\mathcal{A}} \sum_{j=1}^{N_T} \Delta t\, p_j^e \left(\frac{q_{j+1}^e - q_j^e}{\Delta t} - (h_j^e - \psi(q_j^e)) \right) \tag{7.25c}$$

if we set $\lambda_{N_T}^e = p_{N_T}^e = 0$. Assuming sufficient constraint qualifications the first-order optimality system is given by (7.24c) and (7.24b) and the following additional equations for $j \le N_T$, $e \in \mathcal{A}$, and $v \in \mathcal{V}$ and all $\bar{e} \in \mathcal{A}$:

$$\lambda_{j-1}^e = \Delta t \partial_\rho \mathcal{F}(\rho_j^{e,a}, q_j^e) + \lambda_j^e - \frac{\Delta t}{L^e}\left(\phi_j^e - V^e \lambda_j^e\right), \tag{7.26a}$$

$$\phi_j^e := \sum_{\bar{e}\in\delta_v^+ \text{ s.t. } e\in\delta_v^-} p_j^{\bar{e}} \frac{\partial}{\partial \rho^e} h_j^{\bar{e}}, \tag{7.26b}$$

$$p_{j-1}^e = \Delta t \partial_q \mathcal{F}(\rho_j^{e,a}, q_j^e) + p_j^e - \Delta t \left(p_j^e - \lambda_j^e\right)(f^e)'(q_j^e), \tag{7.26c}$$

$$0 = \sum_{e\in\delta_v^+} p_j^e \frac{\partial}{\partial A^{v,\bar{e}}} h_j^e. \tag{7.26d}$$

The summation in the definition of the function ϕ^e is understood in the following way: For a fixed intersection $v \in \mathcal{V}$ such that $e \in \delta_v^-$ we sum over all $\bar{e} \in \delta_v^+$. Hence, the function ϕ^e depends on the type of intersection.

Remark 7.3.4. *We state its explicit form for cases A to C introduced above: In case A we have $\phi_j^2 = 0$ and $\phi_j^1 = p_j^1 V^1$. In case B we obtain $\phi_j^1 = p_j^3 V^1$ and $\phi_j^2 = p_j^3 V^2$. Finally, for the interesting case C we find $e = 1$, which implies $\phi_j^1 = A^{v,2} p_j^2 V^1 + (1 - A^{v,2}) p_j^3 V^1$.*

Furthermore, we obtain with the previous definitions for $\bar{e} \ne \tilde{e}$

$$\sum_{e\in\delta_v^+} p_j^e \partial_{A^{v,\bar{e}}} h_j^e = \left(p_j^{\bar{e}} - p_j^{\tilde{e}}\right) \sum_{e\in\delta_v^-} V^e \rho_j^e. \tag{7.27}$$

Summarizing, the discrete optimality system for (7.24) is given by (7.24b)–(7.24c) and (7.26). Changing the objective functions affects only the first term on the right-hand side in formulas (7.26a) and (7.26c). Again, we can apply a similar algorithm as before by using first the state equations (7.24b)–(7.24c) for a given initial control vector; second we use the adjoint equation (7.26) to solve for the Lagrange multiplier and last the gradient (7.27).

In the following we show that the discrete system (7.24b)–(7.24c) and (7.26) is in fact a suitable discretization of the continuous optimality system. We introduce additional discretizations points on each arc e such that $a, b \to \infty$. Further, we introduce artificial variables $\lambda_j^{e,a} := \lambda_j^e$ and $V^e \lambda_j^{e,b} := \phi_j^e$. Then, the equations read

$$\frac{\rho_{j+1}^{e,b} - \rho_j^{e,b}}{\Delta t} = -\frac{V^e}{L^e}(\rho_j^{e,b} - \rho_j^{e,a}), \quad \rho_0^e = 0,\ V^e \rho_j^{e,a} = f^e(q_j^e), \tag{7.28a}$$

$$\frac{q_{j+1}^e - q_j^e}{\Delta t} = h_j^e - f^e(q_j^e), \quad q_0^e = 0, \tag{7.28b}$$

$$\frac{\lambda_{j-1}^{e,a} - \lambda_j^{e,a}}{\Delta t} = \partial_\rho \mathcal{F}(\rho_j^{e,a}, q_j^e) - \frac{V^e}{L^e}\left(\lambda_j^{e,b} - \lambda_j^{e,a}\right), \quad \lambda_T^{e,a} = 0, \tag{7.28c}$$

$$V^e \lambda_j^{e,b} = \sum_{\bar{e}\in\delta_v^+ \text{ s.t. } e\in\delta_v^-} p_j^{\bar{e}} \frac{\partial}{\partial \rho^e} h_j^{\bar{e}}, \tag{7.28d}$$

$$\frac{p_{j-1}^e - p_j^e}{\Delta t} = \partial_q \mathcal{F}(\rho_j^{e,a}, q_j^e) - \left(p_j^e - \lambda_j^{e,a}\right)(f^e)'(q_j^e), \tag{7.28e}$$

$$0 = \sum_{e \in \delta_v^+} p_j^e \frac{\partial}{\partial A^{v,\bar{e}}} h_j^e. \tag{7.28f}$$

Obviously, (7.28) is an upwind and explicit Euler discretization of (7.29):

$$\partial_t \rho^e + V^e \partial_x \rho^e = 0, \quad \rho^e(x,0) = 0, \quad V^e \rho^e(t,a) = f^e(q^e), \tag{7.29a}$$

$$\partial_t q^e = h^e(\vec{\rho}^e, \vec{A}^v) - f^e(q^e), \quad q^e(0) = 0, \tag{7.29b}$$

$$-\partial_t \lambda^e - V^e \partial_x \lambda^e = \partial_\rho \mathcal{F}(\rho^e(t,a), q^e(t)), \quad \lambda^e(T,x) = 0, \tag{7.29c}$$

$$V^e \lambda^e(t,b) = \sum_{\bar{e} \in \delta_v^+ \text{ s.t. } e \in \delta_v^-} p^{\bar{e}}(t) \frac{\partial}{\partial \rho^{\bar{e}}} h^{\bar{e}}(\vec{\rho}^e, \vec{A}^v), \tag{7.29d}$$

$$-\partial_t p^e = \partial_q \mathcal{F}(\rho^e(t,a), q^e(t)) - \left(p^e - \lambda^e(t,a)\right)(f^e)'(q^e), \quad p^e(T) = 0, \tag{7.29e}$$

$$\sum_{e \in \delta_v^+} p^e \frac{\partial}{\partial A^{v,\bar{e}}} h^e(\vec{\rho}^e, \vec{A}^v) = 0. \tag{7.29f}$$

Therefore, if we formally let $L^e, \Delta t \to 0$ for $L^e/\Delta t$ fixed, we see that λ^e, and furthermore the discrete Lagrangian, tends to the continuous Lagrangian.

In the next section, we propose a reformulation of the discretized versions of problem (7.3) in terms of a mixed-integer model. Depending on the discretization this reformulation allows for an efficient optimization with respect to the distribution at vertices even for large-scale networks.

7.4 Mixed-Integer Programming Model

The applicability of mixed-integer programming in continuous partial differential equation constrained optimization opens a new possibility to solve such problems in an accurate and feasible way.

Before we start introducing the detailed derivation of the discrete optimization problem, we will shortly summarize all important, not only notational, issues on MIPs. For more detailed information we refer the reader to [66, 69]. An MIP model usually consists of a linear cost functional combined with linear constraints which could be simple equations as well as inequalities and different sets of variables. In general, a mixed-integer model can be written as

$$Z(X) = \min\{cx + dy : (x,y) \in X\},$$

where X describes the set of feasible solutions

$$X = \{ (x,y) \in \mathbb{R}_+^n \times \{0,1\}^p : Ax + By \geq C\}.$$

Here, x denotes the n-dimensional vector of nonnegative continuous variables and y denotes a p-dimensional vector of binary variables. The matrices A and B are composed of real coefficients occurring in the constraints, while the right-hand side is $C \in \mathbb{R}^m$. The aim is now to operate on the set of feasible solutions X in order to find an optimal solution for the predefined cost function $cx + dy$. One famous way for solving those MIP models is the Branch-and-Bound algorithm, where subproblems are created by restricting the range of integer variables. In our case, it follows that the variable y is then allowed to take real values in the unit interval. This method is called linear relaxation of the MIP and implies that the feasible set of solutions X is relaxed to

$$X^* = \{\, (x, y) \in \mathbb{R}^n_+, \times [0,1]^p : \; Ax + By \geq C\}.$$

First, starting from the root problem considering X^*, two new subproblems fixing one crucial component of y to 0 or 1, respectively, are analyzed by lower- and upper-bounding procedures. The first step is called branching and the second bounding. Respective lower bounds are provided by the relaxation and compared to the functional value of the root problem for each instance. If both match, a feasible solution has been found; otherwise two further subproblems are generated and the bounding is restarted. There exists a variety of ways to prune the tree of subproblems, but a global optimal solution to the original problem cannot be determined before all nodes have been checked. Due to the complexity of MIP models several heuristics are often required initially and during the solution procedure. The Branch-and-Bound algorithm and its extensions are implemented in standard commercial solvers.

Differently from above we avoid such compact notations since the proposed equations are considered on a time-expanded graph on which they have to satisfy a mixture of coupled equations and inequalities. The procedure is as follows: Starting from the partial differential equation network model (7.3) we derive a simplified model. This is done by using a straightforward two-point discretization of the partial differential equations on each arc of the network. The same principle also holds for the time discretization of the queue. In doing so, the nonlinearity occurring in the queue outflow is approximated by linear equations using binary variables. The main advantages of linear mixed-integer models are the straightforward inclusion of additional constraints (e.g., bounded queues) and the computation of large-scale networks. In this context, we note that other solution methods are also possible. We refer the reader to Laird et al. [60, 61], where networks of realistic size are solved using a nonlinear programming algorithm combined with the origin tracking algorithm. The latter is an alternative discretization for the partial differential equation which also approximates time delays.

7.4.1 Derivation of the MIP

We explain now the direct link to mixed-integer models as a solution technique of continuous optimal control problems such as (7.3). The reformulation as an MIP is based on a coarse grid discretization of the partial differential equation (7.3). This is possible, since (7.3c) does not allow for complex dynamics like backwards travelling shock waves. Hence, we propose a two-point upwind discretization of each arc e. Finally, a reformulation of (7.3e) using binary variables yields the aforementioned mixed-integer model for supply chains. The details are as follows: For each fixed arc $e \in \mathcal{A}$ we introduce two variables for the flux at the boundary and a single variable for the queue for each time j of a time grid $j = 1, \ldots, N_T$:

$$\mathrm{f}_j^{e,a} := f^e(\rho^e(t, a^e)), \quad \mathrm{f}_j^{e,b} = f^e(\rho^e(t, b^e)), \quad \mathrm{q}_j^e := q^e(t) \quad \forall e, j. \tag{7.30}$$

A two-point upwind discretization in space and time of (7.3c) is apparently given by

$$\mathrm{f}_{j+1}^{e,b} = \mathrm{f}_j^{e,b} + \frac{\Delta t}{L^e} V^e \left(\mathrm{f}_j^{e,a} - \mathrm{f}_j^{e,b} \right) \quad \forall e, j, \tag{7.31}$$

where we use the same time discretization Δt for all arcs e. Condition (7.3e) is reformulated by introducing binary variables $\zeta_j^e \in \{0,1\}$ for $e \in \mathcal{A}, j = 1, \dots, N_T$, and is given by

$$\mu^e \zeta_j^e \le \mathrm{f}_j^{e,a} \le \mu^e, \tag{7.32a}$$

$$\frac{\mathrm{q}_j^e}{\epsilon} - \mathcal{M} \zeta_j^e \le \mathrm{f}_j^{e,a} \le \frac{\mathrm{q}_j^e}{\epsilon}, \tag{7.32b}$$

where $\mathcal{M}$ is a sufficiently large constant. To be more precise, $\mathcal{M}$ may be set to $\frac{T}{\epsilon} \max_{e \in \mathcal{A}} \mu^e$.

Further, we need to reformulate the coupling conditions (7.3d) and (7.3e). We introduce variables h_j^e for the total inflow to arc e at $x = a^e$ and require the following equalities for each vertex $v \in \mathcal{V}$:

$$\sum_{e \in \delta_v^+} \mathrm{h}_j^e = \sum_{e \in \delta_v^-} \mathrm{f}_j^{e,b} \quad \forall v, j, \tag{7.33a}$$

$$\mathrm{q}_{j+1}^e = \mathrm{q}_j^e + \Delta t \left(\mathrm{h}_j^e - \mathrm{f}_j^{e,a} \right) \quad \forall e, t. \tag{7.33b}$$

Remark 7.4.1. *The reformulation of the* min *function solely induces binary variables into the problem. All other constraints are linear and can be discretized in a straightforward way. This approach is different from other existing MIP models for supply chains and, furthermore, from a discrete version of the optimality system; cf. Section* 7.3.

Note that we use an explicit time discretization of the ordinary differential equation. This is mainly due to the fact that an implicit discretization would introduce an additional coupling between different arcs on the network. On the contrary, the explicit discretization introduces only a local coupling between the arcs connected at a fixed vertex $v \in \mathcal{V}$. We observe from

$$\partial_t q^e(t) = A^{v,e}(t) \left(\sum_{\tilde{e} \in \delta_v^-} f^{\tilde{e}}(\rho^{\tilde{e}}(t, x_v^{\tilde{e}})) \right) - \min \left\{ \mu^e; \frac{q^e(t)}{\epsilon} \right\}$$

that the ordinary differential equation is stiff, whenever $0 < q^e(t) \le \epsilon \mu^e$. Due to the upwind discretization of the transport equation (7.31) and the relaxation of the queue outflow in the ordinary differential equation (7.33), we need the following fundamental definition to obtain a stable discretization that guarantees correct numerical results.

Definition 7.1.

(a) *The CFL condition for the discretized partial differential equation (see* (7.31)*) reads*

$$\Delta t \frac{V^e}{L^e} \le 1. \tag{7.34}$$

(b) *The so-called stiffness condition for the discretized ordinary differential equation (see* (7.33)*) reads*

$$\Delta t \frac{1}{\epsilon} \leq 1. \tag{7.35}$$

Due to Definition 7.1 we choose the time step size Δt as

$$\Delta t = \min\{\epsilon; L^e / V^e : e \in \mathcal{A}\} \tag{7.36}$$

in the case of the coarsest discretization. A natural choice for ϵ is $\epsilon = \Delta x / V^e$, since, as already mentioned above, $q^e(t)/\epsilon$ represents a relaxed flux. More detailed, we know that a flux can be rewritten as the product of the part density and the processing velocity, i.e., $f^e = V^e \rho^e$. Due to the fact that the density at the first discretization point $x = a^e$ of the processor is the same as $q^e / \Delta x$, the parameter ϵ is determined.

This leads finally to the necessary condition

$$\Delta t = \min\{L^e / V^e : e \in \mathcal{A}\}. \tag{7.37}$$

An analytical interpretation of the CFL condition is shown in Figure 7.4.

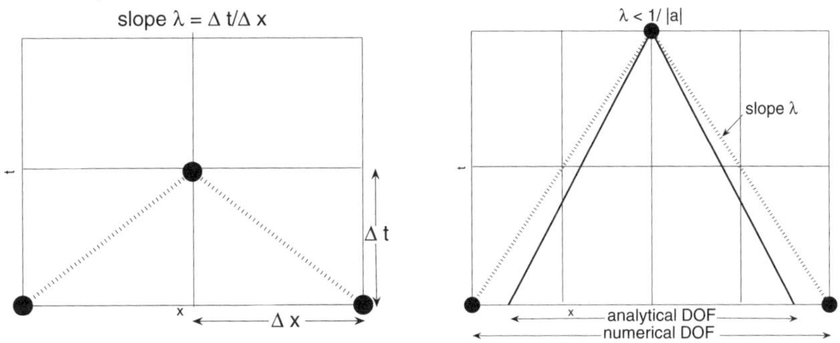

Figure 7.4. *The CFL condition can be traced back to the mathematicians Richard Courant, Kurt Friedrichs, and Hans Lewy in* 1928. *It ensures that the analytical speed of information in a conservation law of the form $\partial_t \rho + a \partial_x \rho = 0$ is slower than the numerical grid velocity defined as λ (left picture); i.e., the analytical domain of dependence is completely covered by the numerical one (right picture). Otherwise, if the condition is injured, numerical instabilities in the solution will occur. For more information we refer the reader to* [63].

Moreover, the optimization problem consists of the following box constraints for all $e \in \mathcal{A}$ and all $j = 1, \dots, N_T$:

$$0 \leq \mathrm{f}_j^{e,a} \leq \mu^e, \quad 0 \leq \mathrm{f}_j^{e,b} \leq \mu^e, \quad 0 \leq \mathrm{q}_j^e. \tag{7.38}$$

Finally, we assign initial data to $\mathrm{f}_1^{e,a}, \mathrm{f}_1^{e,b}$, and q_1^e. For a discretization of the cost functional we use a trapezoidal rule in space and a rectangle rule in time and obtain

$$\sum_{e \in \mathcal{A}} \sum_{j=1}^{N_T} \Delta t \, (L^e/2) \Big(\mathcal{F}(\mathrm{f}_j^{e,a} / V^e, \mathrm{q}_j^e) + \mathcal{F}(\mathrm{f}_j^{e,b} / V^e, \mathrm{q}_t^e) \Big). \tag{7.39}$$

Remark 7.4.2. *Let M denote the number of processors and N_T the number of time intervals. Then, the MIP consists of $3 \cdot M \cdot N_T$ real and $M \cdot N_T$ binary variables.*

Summarizing, the mixed-integer model derived by discretization of the network formulation of the supply chain dynamics is given by

$$\begin{aligned} \min_{A_t^{v,e}, v \in \mathcal{V}_d} \quad & (7.39) \\ \text{subject to} \quad & (7.31), (7.32), (7.33), (7.38). \end{aligned} \tag{7.40}$$

A few remarks are added. First, it is a matter of simple calculations to recover the entries of the distribution vectors $A_t^{v,e}$ from the values of h_j^e. Second, other objective functionals can be envisioned, and in the case of a nonlinearity in (7.1) we might have to introduce additional binary variables to obtain a mixed-integer approximation. This is standard and can be found, for example, in [32]. Third, if we use an implicit discretization of the ordinary differential equation (7.3d),

$$\mathrm{q}_{j+1}^e = \mathrm{q}_j^e + \Delta t \left(\mathrm{h}_{j+1}^e - \mathrm{f}_{j+1}^{e,a} \right), \tag{7.41}$$

we end up with no restriction on the time step. From the continuous point of view such an approach is not favorable due to the additional introduced strong coupling between all arcs in the network. We conclude the modeling with the following remark: In the particular case of a supply chain consisting of a sequence of processors and vertices of degree at most two, there is no possibility to distribute parts. In this case, the mixed-integer model coincides with the two-point upwind discretization of the partial differential equation and both yield the same dynamics. The MIP reduces to a feasibility problem in this case.

Model Extensions

In real-world examples the introduced model is too simple to reflect realistic results. Hence, we propose a few extensions to the mixed-integer model (7.40) on an arbitrary network.

1. **Finite size buffers**

   Usually, in the design of production lines, it is mandatory to limit the size of the buffering queues q_j^e. This condition can be implemented in the mixed-integer context by adding box constraints as follows:

   $$\mathrm{q}_j^e \leq \text{const} \quad \forall e, j. \tag{7.42}$$

   Similarly, we can add the constraints $q^e(t) \leq \text{const}$ to the continuous problem (7.3), which leads to a state-constrained optimal control problem.

2. **Optimal inflow profile**

   Under the assumption of finite sizes in the buffering queues, the question arises to find the maximum possible inflow to the network, such that the buffering capacities of the queues are not exceeded.

This can be modeled by replacing the cost functional (7.39) or (7.1), respectively, by the following objective function:

$$\max \sum_{e \in \mathcal{A}'} \sum_{j=1}^{N_T} \mathrm{f}_j^{e,a}, \tag{7.43}$$

where $\mathcal{A}' \subset \mathcal{A}$ is the set of all inflow arcs of the network.

3. **Processor shutdown due to maintenance**

   Maintenance of processors can also be included in the mixed-integer model: Assume that processor $\tilde{e}$ has to be switched off for maintenance for N consecutive time intervals. Further assume that this period can be chosen freely during the whole simulation time $j = 1, \dots, N_T$. Then, we supplement the mixed-integer model with the following condition:

$$\mathrm{h}_{j+l}^{\tilde{e}} \leq \max\{\mu^e : e \in \mathcal{A}\}|M| \cdot (1 - \phi_j^{\tilde{e}}) \quad \forall j, \forall l = 0, \dots, N-1, \tag{7.44a}$$

$$\sum_{j=1}^{N_T} \phi_j^{\tilde{e}} = 1, \tag{7.44b}$$

   where for each processor $e \in \mathcal{A}$ and every time t we introduce a binary variable $\phi_j^e \in \{0,1\}$ that indicates whether process e is shut down at time j. If $\phi_{j_0}^{\tilde{e}} = 1$, then the maintenance interval starts at time j_0, and in the time interval $j_0, j_0 + N$, the processor $\tilde{e}$ is not available.

Subsequently, we will see that the presented MIP is in fact a linear program (LP) under reasonable assumptions. This result is far from being trivial and requires a detailed numerical analysis of the discrete adjoint equations.

7.5 Reformulation of the MIP as the LP

We show that for a special class of parameter choices in the objective functional the optimal solution of the MIP presented in Section 7.4 and the associated LP are the same. Note that in a pure LP integer variables are not used anymore. This beneficial result is obtained by the application of adjoint equations occurring in the optimality system of the continuous supply network model in Section 7.3 and a duality principle. However, the key idea is to use a different linearization for the same set of constraints. The main advantages of the LP formulation over the MIP is its faster solution time and the possibility to optimize much larger networks.

We start our discussion by briefly recalling some elementary notations which are used throughout this chapter. As already stated, for a fixed vertex v, the set of ingoing arcs is denoted by δ_v^- and the set of outgoing arcs by δ_v^+. Additionally, for a vertex $v \in \mathcal{V}$, we label the arcs $e \in \delta_v^- \cup \delta_v^+ \subset \mathcal{A}$ by $e = 1, \dots, k+l$, where $k = |\delta_v^-|$ and $l = |\delta_v^-|$. For notational convenience, we use $e = 1, \dots, k$ for the ingoing arcs and $e = k+1, \dots, k+l$ for the outgoing arcs. It is sufficient to establish our result for an arbitrary node $v \in V$. Figure 7.5 illustrates the situation.

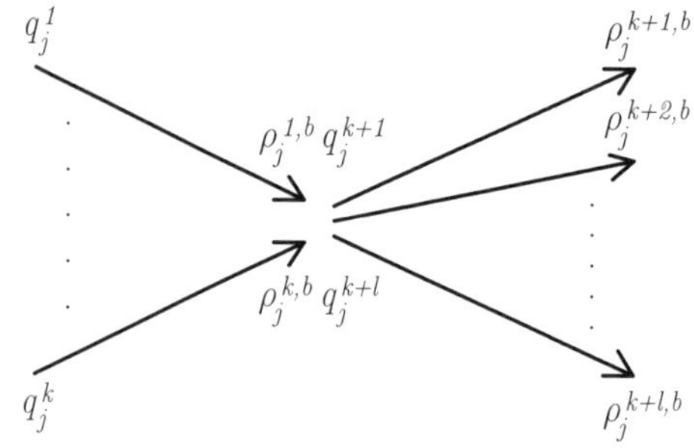

Figure 7.5. *Situation at a node $v \in V$ with k ingoing and l outgoing arcs. Again, q_j^e denotes the value of the queue at time j on arc $e \in \{1,\dots,k+l\}$ and $\rho_j^{e,b}$ describes the content of processor e at time level j at $x = b^e$.*

Before finally stating the discretized supply chain model and the minimization problem we are interested in, we introduce functions h_j^e and ψ_j^e, as already done in Section 7.3, describing the flow distribution at vertices and the exit rates for parts leaving the queue and entering the processor, respectively. For each edge e (resp., $\tilde{e}$) and $\vec{\rho}^e := (\rho^e)_{e\in\mathcal{A}}$ and $\vec{A}^v := (A^{v,e})_{e\in\delta_v^+}$, and fixed $v \in \mathcal{V}$ such that $e \in \delta_v^+$ (resp., $\tilde{e} \in \delta_v^+$), we define

$$h_j^e(\vec{\rho}^e, \vec{A}^v) := A_j^{v,e} \sum_{\bar{e}\in\delta_v^-} V^{\bar{e}} \rho_j^{\bar{e},b} \quad \forall e \in \delta_v^+ \setminus \{\tilde{e}\}, \tag{7.45a}$$

$$h^{\tilde{e}}(\vec{\rho}^e, \vec{A}^v) := \left(1 - \sum_{e\neq\tilde{e}} A_j^{v,e}\right) \sum_{\bar{e}\in\delta_v^-} V^{\bar{e}} \rho_j^{\bar{e},b}. \tag{7.45b}$$

The function $\psi^e(q_j^e)$, originally introduced in [1], is given by

$$\psi^e(q_j^e) = \min\left\{\frac{q_j^e}{\epsilon}, \mu_e\right\}. \tag{7.46}$$

Here, ϵ can be seen as the regularization parameter for the otherwise discontinuous outflow relation. It should be a small positive number; see Remark 7.5.1 below. These definitions are used to write the model in more compact form.

As seen before, the discretized supply chain model is defined on a time grid (subindex j) and a spatial grid. The maximal number of time steps is N_T. The spatial discretization, is essentially a two-point discretization, and, hence, for each fixed arc $e \in \mathcal{A}$ we introduce the variable $\rho_j^{e,b}$ for the density of parts at the boundary b^e. The buffer is described by the variables q_j^e for $j \geq 1$, $e \in \mathcal{A}$. This yields the following dynamics describing the discretized supply chain model:

$$\rho_{j+1}^{e,b} = \rho_j^{e,b} + \frac{\Delta t}{L^e}(\psi^e(q_j^e) - V^e \rho_j^{e,b}), \tag{7.47a}$$
$$q_{j+1}^e = q_j^e + \Delta t(h_j^e - \psi^e(q_j^e)). \tag{7.47b}$$

We introduce an optimization problem by assuming that the controls $A_j^{e,v}$ appearing in the definition of the function h_j^e can be chosen such that the product flow of parts in the network is maximized. Mathematically, this yields the minimization problem

$$\min_{\vec{A}^v,\, v\in\mathcal{V}_d} \sum_{e\in\mathcal{A}} \sum_{j=1}^{N_T} \Delta t \beta_j^e V^e \rho_j^{e,b} \tag{7.48a}$$

$$\text{subject to (7.47a) and (7.47b).} \tag{7.48b}$$

Here, β_j^e are given, negative weights. Finally, we need to prescribe an initial condition for the supply chain and assume for simplicity that the network is initially empty, i.e.,

$$\rho_0^{e,b} = q_0^e = 0 \quad \forall e. \tag{7.49}$$

Remark 7.5.1. *Other choices of the cost functional can be envisioned and used for minimization purposes. The above choice is motivated by the idea to process as many parts as possible in a given time. Therefore, we maximize the flux in each processor.*

The model as given above is only a valid discretization of the corresponding continuous model if and only if two restrictions on the choice of the time step and the regularization parameter are satisfied; cf. Definition 7.1*:*

$$1-\Delta t\frac{V^e}{L^e} \geq 0, \tag{7.50a}$$

$$1-\Delta t\frac{1}{\epsilon} \geq 0. \tag{7.50b}$$

A common way to discretize the equality (7.46) is presented in Section 7.4 using a reformulation in terms of binary variables. Then, the problem (7.48) is a large-scale mixed-integer model; i.e., it is given by

$$\min \sum_{e\in\mathcal{A}} \sum_{j=1}^{N_T} \Delta t \beta_j^e V^e \rho_j^{e,b}, \tag{7.51a}$$

$$\text{subject to } \rho_{j+1}^{e,b} = \rho_j^{e,b} + \frac{\Delta t}{L^e}(\psi^e(q_j^e) - V^e \rho_j^{e,b}), \tag{7.51b}$$

$$q_{j+1}^e = q_j^e + \Delta t(h_j^e - \psi^e(q_j^e)), \tag{7.51c}$$

$$\mu^e \xi_j^e \leq \psi_j^e \leq \mu^e, \tag{7.51d}$$

$$\frac{q_j^e}{\epsilon} - \mathcal{M}\xi_j^e \leq \psi_j^e \leq \frac{q_j^e}{\epsilon}, \tag{7.51e}$$

$$A_j^{v,e} \in [0,1] \tag{7.51f}$$

for $\xi_j^e \in \{0,1\}$ being binary variables and $\mathcal{M}$ being sufficiently large.

However, as seen later in Chapter 8 on numerical results, this problem cannot be solved efficiently for many instances due to the vast number of binary variables. Hence, in the following we proceed differently from now on and show that the problem (7.48), or equivalently (7.51), is in fact an LP.

To proceed, we reformulate the equality condition for (7.46) using a complementarity condition. Introducing the new real variables $u_j^{\pm e}$ we finally obtain an equivalent problem to (7.48). Basically we take advantage of

$$\psi^e(q_j^e) = \min\left\{\mu^e, \frac{q_j^e}{\epsilon}\right\} \Leftrightarrow \psi(q_j^e) = \frac{1}{2}\left(\mu^e + \frac{q_j^e}{\epsilon}\right) - \frac{1}{2}(u_j^{e+} + u_j^{e-}), \tag{7.52a}$$

$$\mu^e - \frac{q_j^e}{\epsilon} = u_j^{e+} - u_j^{e-}, \tag{7.52b}$$

$$u_j^{e+} \cdot u_j^{e-} = 0, \tag{7.52c}$$

$$u_j^{e\pm} \geq 0. \tag{7.52d}$$

In the following section we will prove that there exists indeed an equivalent LP. This will yield an improvement on the computing time for solving the minimization problem that again ensures a global minimum.

7.5.1 An Equivalent Linear Programming Problem

We state the main result before giving the details and interpretations.

Theorem 7.2. *Assume the inflow at a fixed vertex v is not zero, i.e., $\sum_{e\in\delta_v^-} V^e\rho^e > 0$ and $b_{N_T}^e < 0$.* Either *let the inequalities* (7.50) *be strict and $\beta_{j-1}^e \leq \beta_j^e$* or *let $\beta_{j-1}^e < \beta_j^e$ for all j. Then, every optimal solution to the LP*

$$\min \sum_{e\in\mathcal{A}} \sum_{j=1}^{N_T} \Delta t \beta_j^e V^e \rho_j^{e,b} \tag{7.53a}$$

subject to

$$\rho_{j+1}^{e,b} - \left(1 - \Delta t \frac{V^e}{L^e}\right)\rho_j^{e,b} - \frac{1}{2}\frac{\Delta t}{L^e}\mu^e - \frac{1}{2}\frac{\Delta t}{L^e}\frac{q_j^e}{\epsilon} + \frac{1}{2}\frac{\Delta t}{L^e}\left(u_j^{e+} + u_j^{e-}\right) = 0, \tag{7.53b}$$

$$q_{j+1}^e - \left(1 - \frac{1}{2}\frac{\Delta t}{\epsilon}\right)q_j^e - \Delta t h_j^e + \frac{1}{2}\frac{\Delta t}{L^e}\mu^e - \frac{1}{2}\frac{\Delta t}{L^e}\left(u_j^{e+} + u_j^{e-}\right) = 0, \tag{7.53c}$$

$$\mu^e - \frac{q_j^e}{\epsilon} - u_j^{e+} + u_j^{e-} = 0, \tag{7.53d}$$

$$u_j^{e\pm} \geq 0, \tag{7.53e}$$

$$A_j^{v,e} \in [0,1] \tag{7.53f}$$

automatically satisfies $u_j^{e-} \cdot u_j^{e+} = 0$.

Motivation and Interpretation of Theorem 7.2

We offer the following interpretation of Theorem 7.2 and discuss the importance of the assumptions. For simplicity we assume that $L^e = V^e = 1$. The previous equations in (7.53) can be rewritten by utilizing the reformulated queue outflow (7.52a), which is the actual inflow to arc e. We obtain the equivalent system:

$$\rho_{j+1}^{e,b} - \rho_j^{e,b} - \Delta t\left(\psi_j^e - \rho_j^{e,b}\right) = 0, \tag{7.54}$$

$$q_{j+1}^e - q_j^e - \Delta t\left(h_j^e - \psi_j^e\right) = 0, \tag{7.55}$$

$$\psi_j^e + u_j^{e-} = \frac{q_j^e}{\epsilon}, \tag{7.56}$$

$$\psi_j^e + u_j^{e+} = \mu^e, \tag{7.57}$$

$$u_j^{e\pm} \geq 0. \tag{7.58}$$

If we now consider the objective functional (7.53a) and replace $\rho_j^{e,b}$ by (7.54),

$$\sum_{j=1}^{N_T} \beta_j^e \rho_j^{e,b} = \sum_{j=0}^{N_T-1} \beta_j^e \left((1-\Delta t)\rho_{j-1}^{e,b} + \Delta t \psi_j^e\right), \tag{7.59}$$

we observe that maximizing ψ_j^e is advantageous for minimizing the costs at time step j (recall that $\beta_j^e < 0$). Due to the previous equations, maximizing ψ_j^e implies minimizing for either u_j^{e-} or u_j^{e+}. This may serve as an explanation for $u_j^{e+} \cdot u_j^{e-} = 0$ in the optimum, but it is not totally true. In fact, the previous arguments have to be investigated more carefully as the following example shows: Assume the special case $\beta_j^e \equiv -1$, with a time horizon of $N_T = 2$, and consider a single outgoing arc such that

$$h_j = \delta_0(j), \quad \rho_0 = q_0 = 0,$$

where $\delta_0(j)$ denotes the Kronecker delta and the superindex e is skipped. Then, using the first set of equations we have

$$u_0^+ = u_0^- + \mu.$$

According to the time evolution of the queue (7.55) we get

$$q_1 = \Delta t\left(h_0 + u_0^-\right)$$

and, analogously to (7.53d),

$$u_1^+ = u_1^- + \mu - \frac{\Delta t}{\epsilon}\left(h_0 + u_0^-\right).$$

The cost functional is

$$\begin{aligned}
\sum_{j=1}^{N_T} \beta_j^e \rho_j^e &= -\Delta t\left(2\psi_0 + \Delta t \psi_0 - \psi_1\right) \\
&= -\Delta t\left(u_1^- + u_0^-\left(2 - \Delta t - \frac{\Delta t}{\epsilon}\right)\right) + \frac{(\Delta t)^2}{\epsilon} h_0.
\end{aligned}$$

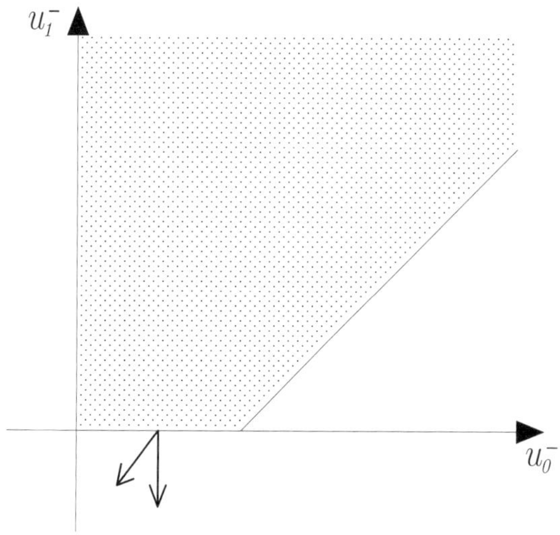

Figure 7.6. *For $\epsilon = \Delta t$ the cost vector is orthogonal to the boundary of the feasible domain given by the set $\{(x,0)|0 \le x \le \frac{\epsilon}{\Delta t}(\mu - \frac{\Delta t}{\epsilon}h_0)\}$. In the case $\epsilon > \Delta t$ the cost vector turns in the depicted direction. This implies that we have a unique optimum.*

Consider the feasible set for u_1^-, u_0^-. Recall that $u_1^+ \ge 0$, and therefore the feasible domain for u_1^- is given by the equation $u_1^- \ge \frac{\Delta t}{\epsilon} u_0^- - (\mu - \frac{\Delta t}{\epsilon} h_0)$. Recall that for $V^e = L^e = 1$ we have $\Delta t = 1$ due to the CFL condition for the partial differential equation. Hence, in the case $\Delta t = \epsilon = 1$ the cost vector is orthogonal to a boundary of the feasible set (see Figure 7.6). Therefore, in this case, the solution is not unique and in particular there exists a solution with $u_0^- \ne 0$. The remedy is to choose $\epsilon > \Delta t$, as stated in the theorem and used in the proof.

7.5.2 Outline of the Proof

The proof of Theorem 7.2 consists of several steps. We therefore first give an overview on the main ideas and techniques before giving the details.

First, we state the KKT system for (7.53). Any optimal solution necessarily has to satisfy these equations. Further, it is sufficient to consider a nodewise analysis of the equations. Second, we prove that any optimal solution to the LP automatically satisfies $u_j^{e+} \cdot u_j^{e-} = 0$. To this end we show that it suffices to prove that (7.62b) and (7.62c) cannot both be equal to zero; see Lemma 7.5. This can be shown, provided that $\frac{1}{L^e}\lambda_j^e - p_j^e > 0$; cf. (7.70) in Lemma 7.4. The last equations are proved by induction on the time variable j.

Derivation of the KKT System

We have inequality constraints on the variables $u_j^{e\pm}$ and $A_j^{v,e}$, while all other constraints are equality constraints. Hence, the KKT system to (7.53) for $j \le N_T$, $e \in \mathcal{A}$, $v \in \mathcal{V}$ is given by the forward equations (7.60), the backward (adjoint) equations (7.61), and the gradient equations (7.62):

$$\rho_{j+1}^{e,b} - \left(1 - \Delta t \frac{V^e}{L^e}\right)\rho_j^{e,b} - \frac{1}{2}\frac{\Delta t}{L^e}\mu^e - \frac{1}{2}\frac{\Delta t}{L^e}\frac{q_j^e}{\epsilon} + \frac{1}{2}\frac{\Delta t}{L^e}\left(u_j^{e+} + u_j^{e-}\right) = 0, \tag{7.60a}$$

$$q_{j+1}^e - \left(1 - \frac{1}{2}\frac{\Delta t}{\epsilon}\right) q_j^e - \Delta t h_j^e + \frac{1}{2}\mu^e - \frac{1}{2}\left(u_j^{e+} + u_j^{e-}\right) = 0, \tag{7.60b}$$

$$\mu^e - \frac{q_j^e}{\epsilon} - u_j^{e+} + u_j^{e-} = 0, \quad u_j^{e\pm} \geq 0, \quad A_j^{v,e} \in [0,1], \tag{7.60c}$$

$$\lambda_{j-1}^e = -\Delta t V^e \beta_j^e + \left(1 - \frac{\Delta t V^e}{L^e}\right)\lambda_j^e + \Delta t \sum_{m=k+1}^{k+l} p_j^m \frac{\partial}{\partial \rho_j^e} h_j^m, \tag{7.61a}$$

$$p_{j-1}^e = \left(1 - \frac{1}{2}\frac{\Delta t}{\epsilon}\right) p_j^e + \frac{1}{2}\frac{\Delta t}{L^e \epsilon}\lambda_j^e + \frac{1}{\epsilon} z_j^e, \tag{7.61b}$$

$$\Delta t (p_j^e - p_j^{k+l}) \sum_{z=1}^{k} V^z \rho_j^z = \begin{cases} = 0, & \text{if } A_j^{v,e} \in (0,1), \\ \geq 0, & \text{if } A_j^{v,e} = 0, \\ \leq 0, & \text{if } A_j^{v,e} = 1, \end{cases} \tag{7.62a}$$

$$\frac{1}{2}\frac{\Delta t}{L^e}\lambda_j^e - \frac{1}{2}\Delta t p_j^e + z_j^e = \begin{cases} = 0, & \text{if } u_j^{e-} > 0, \\ \geq 0, & \text{if } u_j^{e-} = 0, \end{cases} \tag{7.62b}$$

$$\frac{1}{2}\frac{\Delta t}{L^e}\lambda_j^e - \frac{1}{2}\Delta t p_j^e - z_j^e = \begin{cases} = 0, & \text{if } u_j^{e+} > 0, \\ \geq 0, & \text{if } u_j^{e+} = 0. \end{cases} \tag{7.62c}$$

We simplified the notation in the previous system to enhance the readability in the following way: for any fixed vertex v the set of incoming arcs is

$$\delta_v^- := \{1, \ldots, k\},$$

and the set of outgoing arcs is given by

$$\delta_v^+ := \{k+1, \ldots, k+l\}.$$

Below we proceed with a nodewise analysis and will use the same notation throughout the following lemma.

Further, the system has to be supplemented by initial and terminal conditions for state and adjoint variables; i.e., we additionally have

$$\rho_0^{e,b} = q_0^e = 0 \quad \forall e$$

and

$$\lambda_{N_T}^e = -\Delta t\, \beta_{N_T}^e V^e, \quad p_{N_T}^e = 0 \quad \forall e.$$

We have the following remark on the multiplier z_j^e; cf. (7.61b).

Remark 7.5.2. *In order to prove Theorem* 7.2 *we do not need to know the exact value of* z_j^e. *Instead it suffices to have the following estimate obtained by combining* (7.62b) *and* (7.62c)*:*

$$|z_j^e| \leq \frac{1}{2}\Delta t \left(\frac{1}{L^e}\lambda_j^e - p_j^e \right).$$

Next, we discuss the detailed structure of $\frac{\partial}{\partial \rho_j^e} h_j^m$ appearing in the adjoint equation. The following reformulations will be used in the subsequent lemmas.

Evaluation of the Term $\frac{\partial}{\partial \rho_j^e} h_j^m$

For a detailed discussion of ingoing arcs $e = 1, \ldots, k$ (cf. Figure 7.5), we have to investigate the controls $A_j^{v,m}$, $m = k+1, \ldots, k+l-1$, appearing in the adjoint equations. We distinguish three cases for a fixed time step j depending on whether a control is at the boundary or not. We have three cases.

(C1) There are no controls at the boundary: $A_j^{v,m} \in (0,1)$ for all $j, m = k+1, \ldots, k+l-1$. Due to (7.62a) we have $p_j^m = p_j^{k+l}$ for $m = k+1, \ldots, k+l-1$. Therefore, we obtain

$$\sum_{m=k+1}^{k+l} p_j^m \frac{\partial}{\partial \rho_j^e} h_j^m = \sum_{z=k+1}^{k+l} A_j^{v,z} p_j^z V^e \tag{7.63a}$$

$$= \sum_{z=k+1}^{k+l-1} A_j^{v,z} p_j^z V^e + \left(1 - \sum_{z=k+1}^{k+l-1} A_j^{v,z} \right) p_j^{k+l} V^e \tag{7.63b}$$

$$= p_j^{k+l} V^e = p_j^m V^e. \tag{7.63c}$$

(C2) One control is at the upper boundary. This implies that one control is equal to one and all others are zero; i.e., $A_j^{v,m} = 1$ for some m and $A_j^z = 0$ for $z = \{k+1, \ldots, k+l-1\} \setminus \{m\}$. We have

$$p_j^m \leq p_j^{k+l}, \quad \text{if } A_j^{v,m} = 1, \tag{7.64a}$$

$$p_j^z \geq p_j^{k+l}, \quad \text{if } A_j^{v,z} = 0. \tag{7.64b}$$

Consequently, by applying (7.64),

$$\sum_{m=k+1}^{k+l} p_j^m \frac{\partial}{\partial \rho_j^e} h_j^m = \sum_{z=k+1}^{k+l} A_j^{v,z} p_j^z V^e = p_j^m V^e \leq p_j^{k+l} V^e \leq p_j^z V^e. \tag{7.65}$$

(C3) Several controls are at the lower boundary: $A_j^{v,r} = 0$ for some $r \in \{k+1, \ldots, k+l\}$. For ease of notation we assume in the following that $A_j^{v,m} = 0$ for $m \in \{k+1, \ldots, k+s\}$ and $A_j^{v,m} \in (0,1)$ for $m \in \{k+s+1, \ldots, k+l\}$ with $s \leq l-2$ such that there are at least two nonzero controls. With these preliminary remarks we obtain from (7.62a)

$$p_j^m \geq p_j^{k+l} \quad \forall m \in \{k+1,\dots,k+s\}, \tag{7.66a}$$

$$p_j^m = p_j^{k+l} \quad \forall m \in \{k+s+1,\dots,k+l-1\}, \tag{7.66b}$$

and finally

$$\sum_{z=k+1}^{k+l} A_j^{v,z} p_j^z V^e = \sum_{z=k+s+1}^{k+l} A_j^{v,z} p_j^z V^e \overset{(7.66b)}{=} p_j^{k+l} \sum_{z=k+s+1}^{k+l} A_j^{v,z} V^e \tag{7.67a}$$

$$= p_j^{k+l} V^e. \tag{7.67b}$$

Estimates on the Multiplier λ_j^e

The following lemma is the main ingredient for proving Theorem 7.2.

Lemma 7.3. *We require the assumptions of Theorem 7.2 to be satisfied.*
Then, we have for $m = k+1,\dots,k+l$

$$p_j^m - \frac{1}{L^e}\lambda_j^e - \beta_j^e > 0 \quad \forall e \in \{1,\dots,k\} \tag{7.68}$$

and

$$-\frac{1}{L^e}\lambda_j^e - \beta_j^e > 0 \quad \forall e \in \{k+1,\dots,k+l\}.$$

Proof. With the other case being similar we prove only the first case. For $j = N_T$ we have $p_{N_T}^m = 0, m = k+1,\dots,k+l$, and $-\frac{1}{L^e}\lambda_j^e - \beta_j^e = -b_{N_T}^e(1 - \Delta t \frac{V^e}{L^e}) > 0$ due to (7.50).

Let (7.68) hold for some j and some $e \in \{1,\dots,k\}$ as the induction hypothesis.

The proof that this statement also holds for $j-1 \leq j$ is done using the adjoint equations (7.61). We have

$$\begin{aligned} p_{j-1}^m - \frac{1}{L^e}\lambda_{j-1}^e - \beta_{j-1}^e = &\left(1 - \frac{1}{2}\frac{\Delta t}{\epsilon}\right) p_j^m + \frac{1}{2}\frac{\Delta t}{L^m \epsilon}\lambda_j^m + \frac{1}{\epsilon} z_j^m \\ &- \beta_{j-1}^e + \frac{\Delta t V^e}{L^e}\beta_j^e - \left(1 - \frac{\Delta t V^e}{L^e}\right)\frac{1}{L^e}\lambda_j^e \\ &- \frac{\Delta t}{L^e}\sum_{m=k+1}^{k+l} p_j^m \frac{\partial}{\partial \rho_j^e} h_j^m. \end{aligned} \tag{7.69}$$

We use Remark 7.5.2, in particular

$$z_j^m = -\frac{1}{2}\Delta t \left(\frac{1}{L^m}\lambda_j^m - p_j^m\right).$$

Then the detailed discussion in the previous subsections yields a lower bound for (7.69):

$$p_{j-1}^m - \frac{1}{L^e}\lambda_{j-1}^e - \beta_{j-1}^e \geq \left(1 - \Delta t \frac{V^e}{L^e}\right)\left(p_j^m - \frac{1}{L^e}\lambda_j^e - \beta_j^e\right) + \beta_j^e - \beta_{j-1}^e.$$

The summands are positive due to the induction hypothesis and the assumptions in Theorem 7.2. □

Lemma 7.3 enables us to prove that under certain assumptions we can ensure positivity of $\frac{1}{L^e}\lambda_j^e - p_j^e > 0$ on ingoing arcs $e \in \{1,\dots,k\}$.

Lemma 7.4. *Let the assumptions from Theorem 7.2 be satisfied. Then we have*

$$\frac{1}{L^e}\lambda_j^e - p_j^e > 0 \quad \forall e \in \{1,\dots,k+l\}. \tag{7.70}$$

Proof. Again, we prove only the case $e \le k$. The proof is done using induction. For $j = N_T$ the basis is $\frac{1}{L^e}\lambda_{N^T}^e = -\Delta t \beta_{N_T}^e \frac{V^e}{L^e} > 0$ and $p_{N_T}^e = 0$. Let (7.70) hold for some j as the induction hypothesis. We show that the inequality also holds for $j-1 \le j$.

Due to Remark 7.5.2 and the induction hypothesis we obtain a lower bound for $\frac{1}{L^e}\lambda_{j-1}^e - p_{j-1}^e$ if we consider

$$z_j^e = \frac{1}{2}\Delta t\left(\frac{1}{L^e}\lambda_j^e - p_j^e\right).$$

We use the adjoint equations to obtain

$$\begin{aligned}\frac{1}{L^e}\lambda_{j-1}^e - p_{j-1}^e \ge \left(1-\frac{\Delta t}{\epsilon}\right)\left(\frac{1}{L^e}\lambda_j^e - p_j^e\right) - \Delta t\frac{V^e}{L^e}\left(\frac{1}{L^e}\lambda_j^e + \beta_j^e\right)\\ + \frac{\Delta t}{L^e}\sum_{m=k+1}^{k+l} p_j^m \frac{\partial}{\partial \rho_j^e} h_j^m.\end{aligned}$$

By analogous considerations as in the previous proof we obtain the lower bound

$$\frac{1}{L^e}\lambda_{j-1}^e - p_{j-1}^e \ge \left(1-\frac{\Delta t}{\epsilon}\right)\left(\frac{1}{L^e}\lambda_j^e - p_j^e\right) + \Delta t\frac{V^e}{L^e}\left(p_j^m - \frac{1}{L^e}\lambda_j^e - \beta_j^e\right).$$

The terms are positive due to the assumptions in Theorem 7.2, the induction hypothesis, and Lemma 7.3. □

Proof of the Main Theorem

Lemma 7.5. *If*

$$\frac{1}{L_e}\lambda_j^e - p_j^e > 0 \quad \forall e \in \{1,\dots,k+l\},$$

then we have either $u_j^{e-} = 0$ *and* $u_j^{e+} > 0$, or $u_j^{e-} > 0$ *and* $u_j^{e+} = 0$, or $u_j^{e-} = 0$ *and* $u_j^{e+} = 0$.

Proof. Due to the assumption and according to (7.62b) and (7.62c) the case $u_j^{e-} > 0$ and $u_j^{e+} > 0$ cannot occur. □

Finally, by means of the aforementioned results, let us prove Theorem 7.2.

Proof. We prove the assertion only for the more difficult first case; i.e., we assume the condition (7.50b) to be strict and the β_j^e to be monotone but *not* strictly monotone. Then, any optimal solution to the linear program (7.53) necessarily has to satisfy the system (7.60). Due to Lemma 7.4 we have the positivity of $\frac{1}{L^e}\lambda_j^e - p_j^e$ for any arc e and time j. Using Lemma 7.5 we obtain

$$u_j^{e+} \cdot u_j^{e-} = 0.$$

This proves the assertion. □

Summarizing, we have proved that under reasonable assumptions on the cost functional or parameters in the model equations the optimal solutions to the mixed-integer model can be obtained by solving a certain suitable linear programing problem. The crucial point of the proof is the detailed analysis of the first-order optimality system of the LP.

7.6 Summary

As we have seen, important aspects in supply chain decision making are optimization problems such as maximizing output of a production process or minimizing used buffers. To solve the optimization problem, we have proposed two different techniques: adjoint calculus and mixed-integer programming. The former reduces the partial differential equation–ordinary differential equation restricted problem to an unrestricted model where the Lagrange principle can be applied directly. Second, we derived a simplified model by using a straightforward two-point discretization of the equations on each arc. This is strongly connected to the exploration of connections to the description of supply chains by methods of discrete optimization since the resulting equations are interpreted as an MIP. Advantages of such MIP models are the possibility of extensions (e.g., bounded queues) whose inclusion into the Lagrange principle would lead to a state-constrained problem and a completely new calculus as well as numerical implementation. We have also proved the equivalence of an LP, an MIP, and a complementarity problem arising in the context of flow networks. The proof is based on a detailed analysis of the first-order optimality system of the LP.

Chapter 8

Computational Results

8.1 Simulation

8.1.1 Comparison to Discrete Event Simulation

This section is devoted to the presentation of some computational experiments validating the models introduced in Chapters 3 and 5. We show results which point out the main differences between the model of Armbruster, Degond, and Ringhofer (5.1) and the network model (5.9). A comparison of computing times underlines the computational benefit of using partial differential equation models for large-scale networks. We start by presenting numerical results for the simulation and validation of the different supply chain models. Further, we emphasize differences and correlations of the models (5.1) and (5.9), for instance installing queues to avoid delta distributions and cutting off the flux by the min function. Numerically, we discretize the system (5.9) using the upwind scheme for the partial differential equations and an explicit Euler scheme for the ordinary differential equations. Therein, each arc e could have different space increments, namely $\Delta x^e = L^e/D^e$, where $D^e = 10$ is the number of space discretization points for one processor. For simplicity, the time steps Δt are constant and satisfy the CFL condition on each arc.

Definition 8.1 (CFL condition for networks). *The CFL condition guarantees the stability of any explicit discretization scheme by time step restriction*

$$\Delta t \leq \min_{e \in \mathcal{A}} \left\{ \frac{\Delta x^e}{V^e} \right\}, \tag{8.1}$$

where V^e denotes the processing velocity used in the advection equation for the processors.

Our computations are performed on a linear network; i.e., we consider a supply chain consisting of four processors characterized by L^e, μ^e, and T^e for $e = 1, \ldots, 4$; see the data in Table 8.1. We use the labeling according to our network definition (Definition 5.1).

Initial values for the first arc are fixed by the inflow profile $f^{in}(t)$; see Figure 8.1 and the discussion in Remark 5.2.1. All other initial values are zero, i.e., $\rho^{e,0} = 0$ and $q^{e,0} = 0$; i.e., we start with an empty network. The time horizon is $T = 200$. Note that in this setting

Table 8.1. *Parameters of the sample problem.*

Processor e	μ^e	T^e	L^e
1	14	1	1
2	10	1	1
3	15	1	1
4	9	1	1

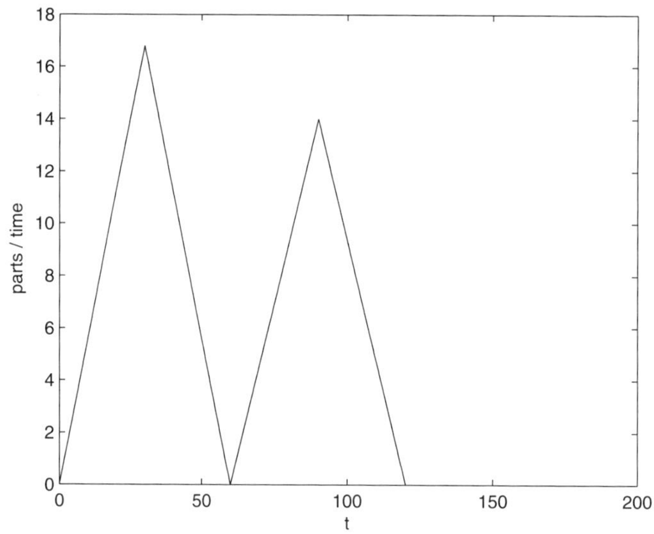

Figure 8.1. *Inflow profile $f^{in}(t)$ prescribed as initial data on an artificial first arc.*

bottlenecks occur at processors 1, 2, and 4. A bottleneck is a situation where the incoming flux into one processor is higher than its maximal capacity.

In Figures 8.2 and 8.3 we present the numerical solution to the network model (5.9), whereas Figures 8.4 and 8.5 show the solution of (5.1). At the top, in Figure 8.2, we find plots of the queues which are located in front of the processors. The queues provide buffering areas where the incoming flux exceeding the maximal capacity is stored. Since the influx $f^{in}(t)$ exceeds the maximal capacity of the first processor $e = 1$ for a small time, one peak of the flux is reduced to μ^1 and the surplus is stored in queue 1. Due to the smaller capacity of processor 2, the flux is reduced to this value and queue 2 starts to increase for the times where the incoming flux f^1 is greater than the maximal capacity μ^2. The queue q^3 remains empty, since $\mu^3 > f^2$ for all times. Hence, the flux is simply propagated. In the last processor of the chain, we observe, similar to the bottleneck in processor 2, a filling of the appropriate queue.

The growth of queues is comparable with the development of delta function concentrations in the density ρ; cf. Figure 8.4. In Figure 8.3, we find a plot of the final density

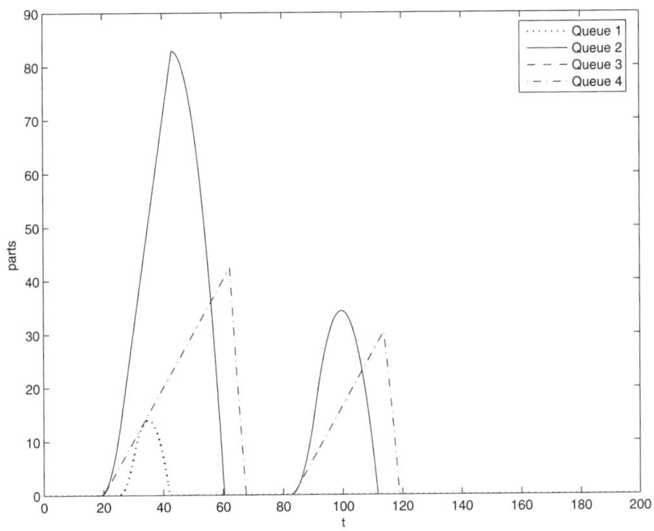

Figure 8.2. *Evolution of the queues in the network model* (5.9). *In queues q^1, q^2, and q^4 we observe the buffering of an exceeding demand.*

of the network model. The density of the first processor $\rho^1(t,x)$ corresponds to the strip $t \geq 0, 0 \leq x \leq 1$, the density for the second processor to $t \geq 0, 1 \leq x \leq 2$, and so forth. Due to the relation $f^e = V^e \rho^e$, where $V^e = 1$ for all e, the density coincides with the network flux. Furthermore, we observe functions that are also plateau shaped since, as explained above, the delta distributions are absorbed by the queues.

To compare the network model with (5.1), we also give a plot of the antiderivative $u(t,x)$ obtained by the calculations in Section 3.3; i.e., we formally integrate $\rho = -\partial_x u$ with respect to x. This technique is used to reformulate the concept of solutions since delta distributions are prevented. As expected, one observes the appearance of sharp peaks approximating delta distributions at the first spatial gridpoint of each processor, here at $x = 0, x = 1$, and $x = 3$. A plot is given in Figure 8.4. Those peaks do not occur in the density of the network model (Figure 8.3), due to the introduced buffering by queues.

8.1.2 Computing Times

We give results on the computational effort for solving the partial differential equation network model (5.9) in comparison with the discrete model (3.4)–(3.7), the so-called discrete event simulation. We report on computing times for a sequence of processors as shown in Figure 8.6. For each arc $e \in \mathcal{A}$ the capacities μ^e, velocities V^e, and lengths L^e are all equally set to 1. We use a constant inflow profile at $x = a^1$ and zero initial data on all other arcs. Note that the required number of parts N for the discrete event simulation (3.4) is given by the integration of the inflow profile f^{in} prescribed at $x = a^1$ on arc one and time $T = 200$:

$$N = \int_0^T f^{in}(t)\, dt. \tag{8.2}$$

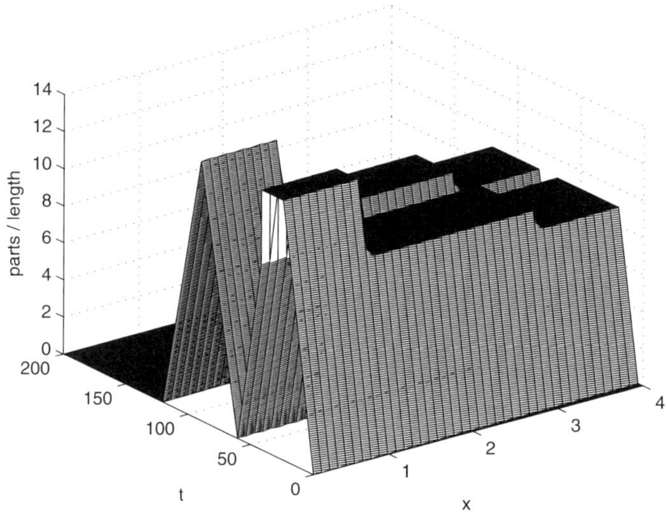

Figure 8.3. *Density of the network model* (5.9). *Except for the density of the third processor* ρ^3*, all other densities are bounded by the maximal capacities.*

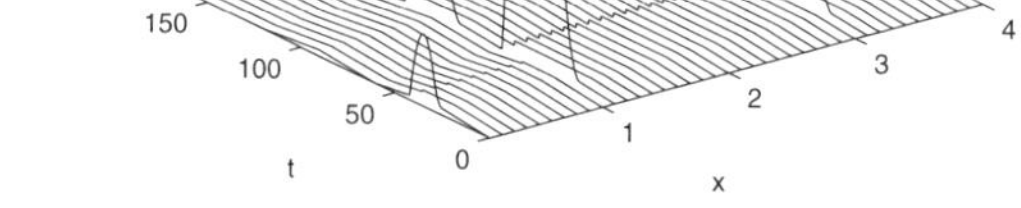

Figure 8.4. *Density with delta function concentrations computed by* (5.1).

Since we are interested in the computation of large quantity production, we vary the number of parts as well as the number of processors. To interpret the results accurately, we point out some technical details for both models. Due to the fact that macroscopic models are independent of a special number of parts, we are able to compare the computing times of the discrete model with the partial differential equation model. There is no difference in the

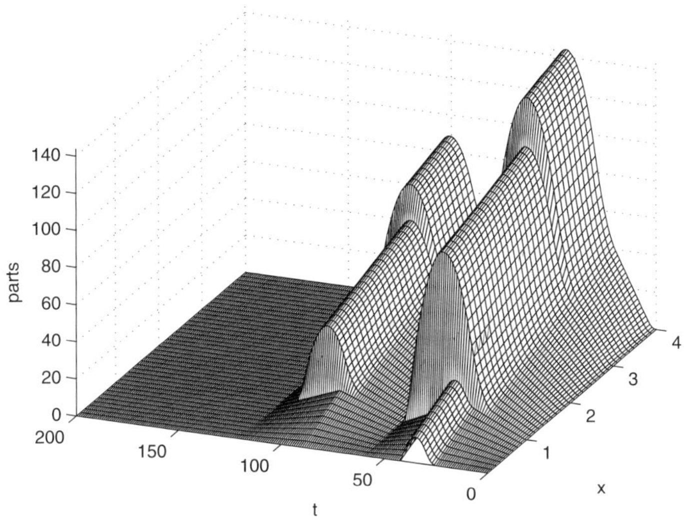

Figure 8.5. *Antiderivative of the density; see* (3.17).

quality of solutions. In contrast to the discrete model, which depends mainly on the number of parts and processors, the computational effort for solving the network model (5.9) is dependent on time and space discretizations. As a coarsest discretization, we propose two discretization points in space for each processor. This corresponds to $N_T = 400$ time steps satisfying the CFL condition; see Definition 8.1.

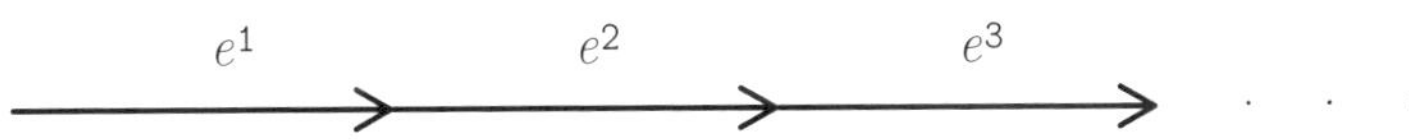

Figure 8.6. *A production network of M processors represented as arcs.*

In Tables 8.2 and 8.3 we present results for different time discretizations, namely a coarse $\Delta t = 0.5$ and a fine $\Delta t = 0.1$ discretization.

The total computation time of the discrete model (3.4)–(3.7) is separated in three different evaluations. Hence, we comment on how time consuming the individual steps are:

1. At first, we determine initial conditions (3.5a) via linear interpolation and, further, the computation of (3.5b). For this purpose, up to 8% of the total time is taken.

2. In a next step, some time (< 1%) is used to compute the matrix $(\tau(m,n))$ in (3.4).

3. The evaluation of (3.4) by N-curves (3.7) takes up the most computing time. Here, more than 90% of the total time is consumed.

In Table 8.2, we observe that solving the partial differential equation model is more efficient in all settings. This is particularly highlighted for $M = 100$ processors, where the partial differential equation model is solved just under one minute and the discrete model needs more than 23 minutes. But as already mentioned, the quality of solutions is the same for large values of the quantity N.

Table 8.2. *CPU times for M processors and* $N_T = 400$ *time intervals.*

Model	Parameters		CPU time
DES	$M = 10$	$N = 20000$	8.8427 sec
DES		$N = 200000$	1.5574 min
PDE			5.558 sec
DES	$M = 50$	$N = 20000$	57.1822 sec
DES		$N = 200000$	11.9164 min
PDE			28.0203 sec
DES	$M = 100$	$N = 20000$	2.0204 min
DES		$N = 200000$	23.6543 min
PDE			57.0921 sec

In Table 8.3, we present results for finer time discretization. It can be seen that in the case of smaller instances, i.e., ($N = 20000$) parts, solving the discrete model is more efficient. For this fine discretization solving the partial differential equation model is comparable with computing 100 000 parts for the discrete event simulation. We recognize further that the CPU times grow linearly with the number of parts in the discrete model.

Table 8.3. *CPU times for M processors and* $N_T = 2000$ *time intervals.*

Model	Parameters		CPU time
DES	$M = 10$	$N = 20000$	27.6397 sec
DES		$N = 200000$	5.1837 min
PDE			2.3161 min
DES	$M = 50$	$N = 20000$	4.5405 min
DES		$N = 200000$	57.492 min
PDE			11.7541 min
DES	$M = 100$	$N = 20000$	10.0446 min
DES		$N = 200000$	114.28 min
PDE			23.5931 min

8.1.3 Priority-Dependent Network Simulation

Concluding this section, we present numerical results for the two-phase ($k = 1,2$) network model introduced in Section 5.5. For simplicity, we restrict our analysis to an example consisting of two processors only. The system of partial differential equations is discretized by applying the Lax–Friedrichs method with 400 gridpoints for each processor. For the discretization of the queues we use the explicit Euler method. Moreover, as initial conditions we assume an empty network. Note that, alternatively, an appropriate discrete event simulation might be performed. Hence, the arrival times τ must be additionally supplemented by attributes which may vary in time. Nevertheless, due to the high computational effort, this ansatz has not been pursued so far.

We consider a situation where the relation between high and low priority flows changes. Let us suppose, for instance, that $Y_1 < Y_2$; then parts labeled by Y_1 are called "high priority" parts and parts labeled by Y_2 are "low priority" parts (due to the fact that lower values of Y_k, $k = 1,2$, correspond to higher priorities).

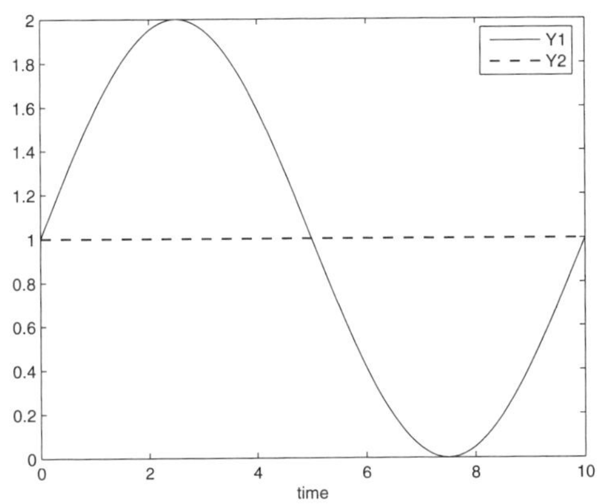

Figure 8.7. *Evolution of the priorities $Y_1(t,0)$ and $Y_2(t,0)$.*

In Figure 8.7, a purely academic scenario, we notice that in the first half of the simulation, as long as $t \leq 5$ holds, priority Y_2 defines the high priority parts (hot lots) which have to be produced first. Then, at $t > 5$, the relation turns over and Y_2 now defines the low priority parts and Y_1 the high priority parts. Furthermore, we assume the processing velocities to be constant. Data for the processors are given by the processing velocities $v_k^e = 3$ and the maximal processing rates $\mu^1 = 2$ and $\mu^2 = 1.5$. The inflow profiles (i.e., the flow concerning each attribute feeding into the queue of processor 1) are defined by the constant function $q_1^{in}(t,0) = 2$ and the piecewise linear decreasing function $q_2^{in}(t,0) = -0.75t + 3$ for $t \leq 4$ in both cases and $q_k^{in} = 0$ otherwise.

The queues are plotted in Figure 8.8. We observe a strong increasing queue π_1^1, while priority Y_2 has higher priority. Since the maximal processing rate is given by $\mu^1 = 2$ there are also parts stored in queue π_2^1. Due to the different processing rates both queues of the second processor are filled. By comparing the queues corresponding to the first attribute we notice lower values of π_1^2 since priority Y_1 defines the hot lot.

In Figure 8.9 the movement of the stack pointers $\mathcal{Y}^v(t)$ is shown. For a better illustration, the case $\mathcal{Y}^v = +\infty$ is displaced by $\mathcal{Y}^v = 10$. The movement of the first pointer can

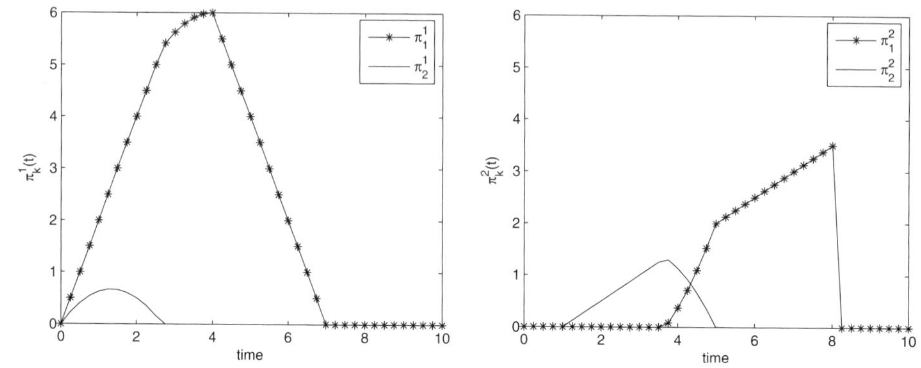

Figure 8.8. *Number of parts in the queues for processor* 1 *(left) and* 2 *(right).*

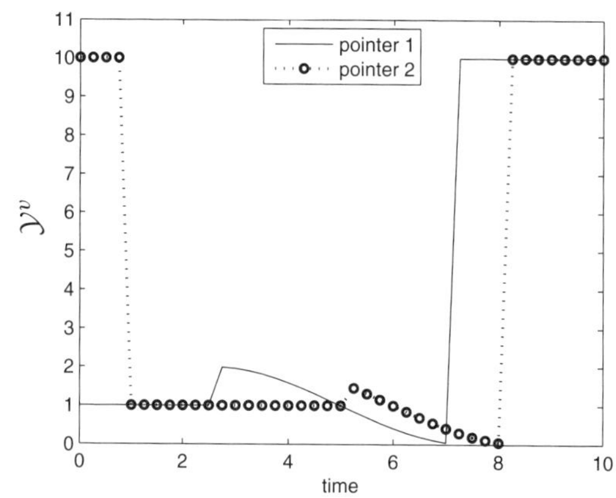

Figure 8.9. *Movement of the pointers* $\mathcal{Y}^v$ *for* $v = 1,2$.

be particularly explained by the definition of the pointer, while the movement of the second pointer completely depends on the dynamics of the priorities Y_k.

8.1.4 Continuum-Discrete Model Simulations and Comparisons

Approximation Algorithm: Fast Godunov for 2 x 2 Systems

Consider the numerical flux function $F(U)$ with $f(\rho,\mu)$ defined in (6.3). Since we want to determine a simplified expression for the numerical flux of the Godunov scheme, we solve Riemann problems between the two states: (ρ_-,μ_-) on the left and (ρ_+,μ_+) on the right. In particular, referring to relation (2.61) we compute the value of the flux function F in the separation point between waves of different speeds.

Theorem 8.2. *The numerical flux function* $G(V,W) = F(U_R(0;V,W))$ *is*

$$G(\rho_-,\mu_-,\rho_+,\mu_+) = \begin{cases} (\rho_-, -\mu_+), & \text{if } \rho_- < \mu_- \vee \rho_- \leq \mu_+, \\ \left(\frac{1-\varepsilon}{1+\varepsilon}\mu_+ + \frac{2\varepsilon}{1+\varepsilon}\rho_-, -\mu_+\right), & \text{if } \rho_- < \mu_- \vee \rho_- > \mu_+, \\ \left(\frac{1+\varepsilon}{2}\rho_- + \frac{1-\varepsilon}{2}\mu_-, -\mu_+\right), & \text{if } \rho_- \geq \mu_- \vee \mu_+ > \tilde{\mu}, \\ \left(\frac{1-\varepsilon}{1+\varepsilon}(\mu_+ + \varepsilon\mu_-) + \varepsilon\rho_-, -\mu_+\right), & \text{if } \rho_- \geq \mu_- \vee \mu_+ \leq \tilde{\mu}, \end{cases} \tag{8.3}$$

with

$$\tilde{\mu} = \mu_- + \frac{1+\varepsilon}{2}(\rho_- - \mu_-). \tag{8.4}$$

Proof. Let P be the intersection point between the first family curve passing through (ρ_-, μ_-) and the line $\rho = \mu$, namely $P = \binom{\rho_-}{\rho_-}$. The second family curve passing through P splits the invariant region into two regions $A = \{(\rho,\mu) : \mu > \rho_-\}$ (in gray) and $B = \{(\rho,\mu) : \mu \leq \rho_-\}$ as depicted in Figures 8.11 and 8.12.

Each Riemann problem solution presents waves travelling with two velocities, namely $\lambda_0 = -1$ and $0 < \varepsilon \leq \lambda_1 \leq 1$. Let (ρ_*, μ_*) be the intermediate state; see Figure 8.10. We compute the numerical flux function $G(\rho_-, \rho_+)$ given by $(f(\rho_*, \mu_*), \mu_*)$. We distinguish two cases.

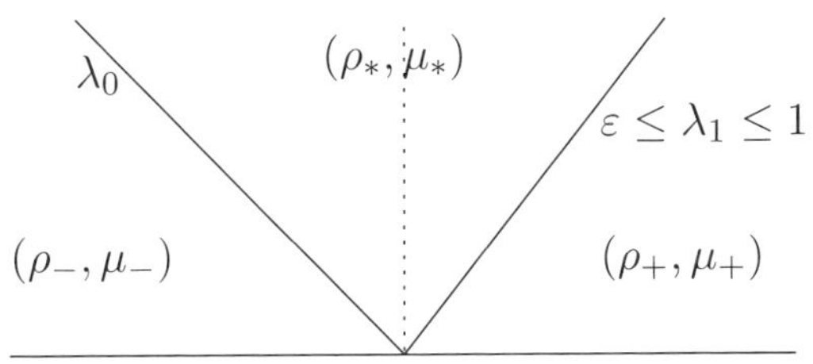

Figure 8.10. *Intermediate state between the two waves.*

Case 1 $\rho_- < \mu_-$;

Case 2 $\rho_- \geq \mu_-$.

In Case 1, if $(\rho_+, \mu_+) \in A$, then $(\rho_*, \mu_*) = (\rho_-, \mu_+)$. Since $\rho_- \leq \mu_+$ the flux (6.4) results to be $f(\rho_-, \mu_+) = \rho_-$.
If $(\rho_+, \mu_+) \in B$, the needed value of the flux is that corresponding to $(f(\rho_*, \mu_+), -\mu_+)$; see Figure 8.11.

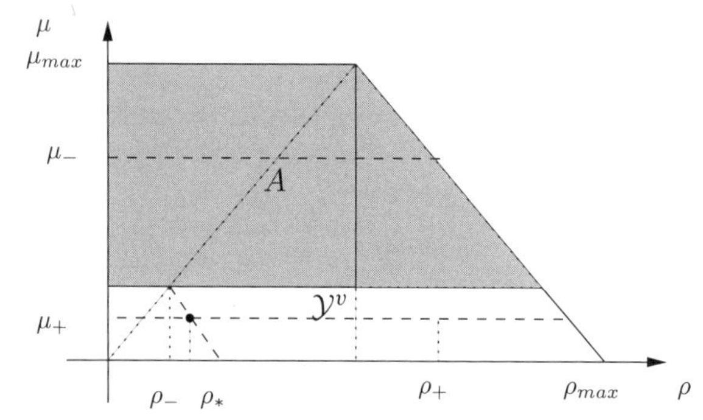

Figure 8.11. *Case* 1, *with* $(\rho_+,\mu_+) \in B$.

We have

$$(\rho_*,\mu_*) = (\rho_*,\mu_+) = \begin{pmatrix} \rho_- \\ \rho_- \end{pmatrix} + t \begin{pmatrix} -\frac{1-\varepsilon}{1+\varepsilon} \\ 1 \end{pmatrix}, \tag{8.5}$$

and ρ_* is computed as

$$\rho_* = \rho_- + (\rho_- - \mu_+)\frac{1-\varepsilon}{1+\varepsilon}. \tag{8.6}$$

Finally, since $\rho_* > \rho_- > \mu_+$ we get the expression in the second line of (8.3). In Case 2, if $(\rho_+,\mu_+) \in A$ (see Figure 8.12), then $(\rho_*,\mu_*) = (\tilde{\rho},\mu_+)$, where

$$\tilde{\rho} = \frac{1+\varepsilon}{2}\rho_- + \frac{1-\varepsilon}{2}\mu_- \tag{8.7}$$

is obtained as follows. The point $(\tilde{\rho},\tilde{\mu})$ is

$$(\tilde{\rho},\tilde{\mu}) = \begin{pmatrix} \rho_- \\ \mu_- \end{pmatrix} + t \begin{pmatrix} -\frac{1-\varepsilon}{1+\varepsilon} \\ 1 \end{pmatrix},$$

and, using that $\tilde{\rho} = \tilde{\mu}$, one gets (8.7). Assuming $(\rho_+,\mu_+) \in B$, the value of the flux we need is $f(\rho_*,\mu_+)$ with ρ_* given by

$$(\rho_*,\mu_*) = (\rho_*,\mu_+) = \begin{pmatrix} \rho_- \\ \mu_- \end{pmatrix} + t \begin{pmatrix} -\frac{1-\varepsilon}{1+\varepsilon} \\ 1 \end{pmatrix}, \tag{8.8}$$

and, making simple computations, one gets

$$\rho_* = \rho_- + (\mu_- - \mu_+)\frac{1-\varepsilon}{1+\varepsilon}. \tag{8.9}$$

Taking into account that $\rho_* > \mu_+$, we obtain the expression of flux as in the last line of (8.3). □

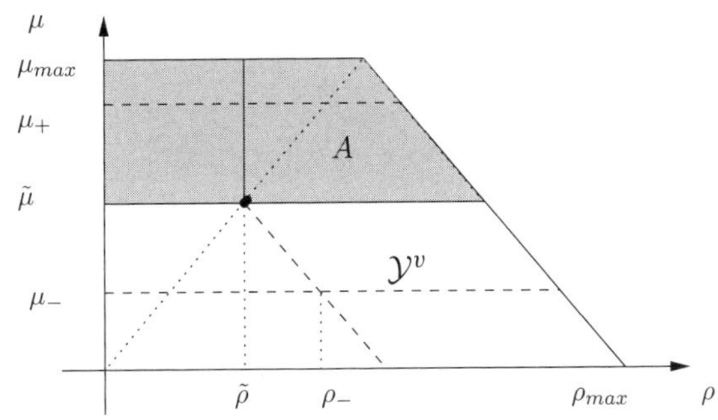

Figure 8.12. *Case* 2, *with* $(\rho_+, \mu_+) \in A$.

Numerics for Riemann Solvers

In this section we describe the numerical framework for the solution of Riemann problems at junctions. In particular, we refer to the general Riemann Solver called SC1, already proposed in [26] and described in Section 6.5, and to the Riemann Solvers introduced in Sections 6.7 and 6.8, namely SC2 and SC3.

For simplicity, we focus on a single node v^e and two consecutive arcs, namely $e, e+1$. Let us introduce the notations:

- $\rho_n^{e,L}, \mu_n^{e,L}$ are the approximate values, respectively, of the density and processing rate at time t_n at the outgoing endpoint $x_L = L\Delta x$ of arc e;
- $\rho_n^{e,0}, \mu_n^{e,0}$ are the approximate values, respectively, of the density and processing rate at time t_n at the incoming endpoint $x_0 = 0$ of arc $e+1$.

Let us now describe the discretization of the Riemann Solver SC1. If we set

- $\hat{\gamma} = f(\rho_n^{e,L}, \mu_n^{e,L})$,
- $\gamma_{\max}^{e+1} = f(\rho_{\max}^{e+1}, \mu_n^{e+1,0})$,

we have two cases.

Case α If $\hat{\gamma} \le \gamma_{\max}^{e+1}$, then

$$\rho_n^{e,L+1} = \rho_n^{e,L},$$

$$\mu_n^{e,L+1} = \mu_n^{e,L},$$

$$\rho_n^{e+1,-1} = \begin{cases} f(\rho_n^{e,L}, \mu_n^{e,L}), & \text{if } f(\rho_n^{e,L}, \mu_n^{e,L}) \le \mu_n^{e+1,0}, \\ \frac{f(\rho_n^{e,L}, \mu_n^{e,L}) - \mu_n^{e+1,0}}{\varepsilon} + \mu_n^{e+1,0}, & \text{otherwise}, \end{cases}$$

$$\mu_n^{e+1,-1} = \mu_n^{e+1,0}.$$

Case β If $\hat{\gamma} > \gamma_{\max}^{e+1}$, then

$$\begin{aligned}
\rho_n^{e,L+1} &= \rho_n^{e,L},\\
\mu_n^{e,L+1} &= \frac{\gamma_{\max}^{e+1} - \varepsilon\rho_n^{e,L}}{1-\varepsilon},\\
\rho_n^{e+1,-1} &= \rho_{\max}^{e+1},\\
\mu_n^{e+1,-1} &= \mu_n^{e+1,0}.
\end{aligned}$$

The discretized version of the Riemann Solver SC2 is given below.

Case α We distinguish between the following subcases:

(α_1) If $\rho^* < \rho_M$, we set

$$\begin{aligned}
\rho_n^{e,L+1} &= \rho_n^{e,L},\\
\mu_n^{e,L+1} &= \min\{\rho^*, \mu_{\max}^{e}\},\\
\rho_n^{e+1,-1} &= \rho^*,\\
\mu_n^{e+1,-1} &= \mu_n^{e+1,0};
\end{aligned}$$

(α_2) if $\rho^* \geq \rho_M$, the new values are

$$\begin{aligned}
\rho_n^{e,L+1} &= \rho_n^{e,L},\\
\mu_n^{e,L+1} &= \varepsilon\frac{1+\varepsilon}{1-\varepsilon}\tilde{\rho} - \frac{2\varepsilon}{1-\varepsilon}\bar{\mu}^e + (1+\varepsilon)\mu_n^{e+1,0},\\
\rho_n^{e+1,-1} &= \tilde{\rho},\\
\mu_n^{e+1,-1} &= \mu_n^{e+1,0};
\end{aligned}$$

Case β

$$\begin{aligned}
\rho_n^{e,L+1} &= \rho_n^{e,L},\\
\mu_n^{e,L+1} &= \bar{\mu}^e,\\
\rho_n^{e+1,-1} &= \bar{\mu}^e,\\
\mu_n^{e+1,-1} &= \mu_n^{e+1,0}.
\end{aligned}$$

The discretization of the Riemann Solver SC3 is described in what follows.

Case α The following subcases can occur:

(α_1) if $\rho^* < \rho_M$, we set

$$\begin{aligned}
\rho_n^{e,L+1} &= \rho_n^{e,L},\\
\mu_n^{e,L+1} &= \max\{\rho^*, \mu_n^{e,L}\},\\
\rho_n^{e+1,-1} &= \rho^*,\\
\mu_n^{e+1,-1} &= \mu_n^{e+1,0};
\end{aligned}$$

(α_2) if $\rho^* \geq \rho_M$, we compute the new values as in SC2:

$$\rho_n^{e,L+1} = \rho_n^{e,L},$$

$$\mu_n^{e,L+1} = \varepsilon\frac{1+\varepsilon}{1-\varepsilon}\tilde{\rho} - \frac{2\varepsilon}{1-\varepsilon}\bar{\mu}^e + (1+\varepsilon)\mu_n^{e+1,0},$$

$$\rho_n^{e+1,-1} = \tilde{\rho},$$

$$\mu_n^{e+1,-1} = \mu_n^{e+1,0};$$

Case β

$$\rho_n^{e,L+1} = \rho_n^{e,L};$$

(β_1) if $\mu_n^{e,L+1} \geq \bar{\mu}^e$, we set

$$\mu_n^{e,L+1} = \mu_n^{e,L};$$

(β_2) otherwise, we assign

$$\mu_n^{e,L+1} = \bar{\mu}^e,$$

$$\rho_n^{e+1,-1} = \bar{\mu}^e,$$

$$\mu_n^{e+1,-1} = \mu_n^{e+1,0}.$$

Numerical Tests

As an application of the supply chain dynamics presented in Chapter 6 and the associated numerical algorithm described here, we present some experiments on sample cases. The problem (6.3) is discretized using the Godunov and upwind schemes, as indicated in (2.63), with the numerical flux (8.3). We set an equal space increment on each arc, namely $N^e = \frac{L^e}{\Delta x}$, where N^e is the number of space discretization points. The time steps Δt are constants and are obtained imposing the CFL condition on each arc.

In the following tests (Tests 1 and 2) we refer the reader to numerical examples presented in [37, 38], in such a way to establish a comparison between their approach and ours. To this aim we consider the flux function with different slopes (6.5). The expression of numerical flux $G(\rho_-, \mu_-, \rho_+, \mu_+)$ of the fast Godunov scheme for arc e is

$$G = \begin{cases} (m^e\rho_-, -\mu_+), & \text{if } \rho_- < \mu_- \vee \rho_- \leq \mu_+, \\ \left(\left(m^e - \frac{2\varepsilon}{1+\varepsilon}\right)\mu_+ + \frac{2\varepsilon}{1+\varepsilon}\rho_-, -\mu_+\right), & \text{if } \rho_- < \mu_- \vee \rho_- > \mu_+, \\ \left(m^e\left(\frac{1+\varepsilon}{2}\rho_- + \frac{1-\varepsilon}{2}\mu_-\right), -\mu_+\right), & \text{if } \rho_- \geq \mu_- \vee \mu_+ > \tilde{\mu}, \\ \left(\left(m^e - \frac{2\varepsilon}{1+\varepsilon}\right)\mu_+ + \frac{\varepsilon(1-\varepsilon)}{1+\varepsilon}\mu_- + \varepsilon\rho_-, -\mu_+\right), & \text{if } \rho_- \geq \mu_- \vee \mu_+ \leq \tilde{\mu}, \end{cases}$$

with $\tilde{\mu}$ as in (8.4).

Test 1. As in [38], we consider a supply chain network consisting of $N = 4$ arcs and we use the data in Table 8.4.

Table 8.4. *Parameters of Test* 1.

Processor e	μ^e	m^e	L^e
1	25	1	1
2	15	0.2	0.2
3	10	0.2	0.6
4	15	0.2	0.2

Let us assume the following initial and boundary data:

$$\rho_1(0,x) = \rho_2(0,x) = \rho_3(0,x) = \rho_4(0,x) = 0,$$

$$\rho_1(t,0) = \begin{cases} \frac{18}{35}t, & \text{if } 0 \le t \le 35, \\ -\frac{18}{35}t + 36, & \text{if } 35 < t \le 70, \\ 0, & \text{if } t > 70, \end{cases}$$

and the space and time intervals are, respectively, $[0,2]$ and $[0,140]$, with $\Delta x = 0.02$ and $\Delta t = 0.01$. On each processor $e = 1,2,3,4$ we assume as the initial datum $\mu(0,x)$ the value μ^e, which is also imposed at the incoming and outgoing boundaries. Notice that the inflow profile $\rho_1(t,0)$ is assigned on the first processor, which can be considered as an artificial arc, and it exceeds the maximum capacity of the other processors.

In Figure 8.13 we depict the evolution in time on processors 2, 3, 4 of the flux, density, and processing rate, obtained by the Riemann Solver SC1 for $\varepsilon = 0.1$.

From the analysis of graphics in Figure 8.13, we can deduce that the processing rate, according to SC1, is minimized and, consequently, the flux and the density are considerably lowered and are almost plateau shaped on processors 3 and 4. On the other hand, SC2 determines the behavior shown in Figure 8.14, where the flux and the density are correctly developed on processors 2, 3, 4, due to the behavior of the processing rate depicted in the graphics, which assumes the minimum possible value in order to maximize the flux.

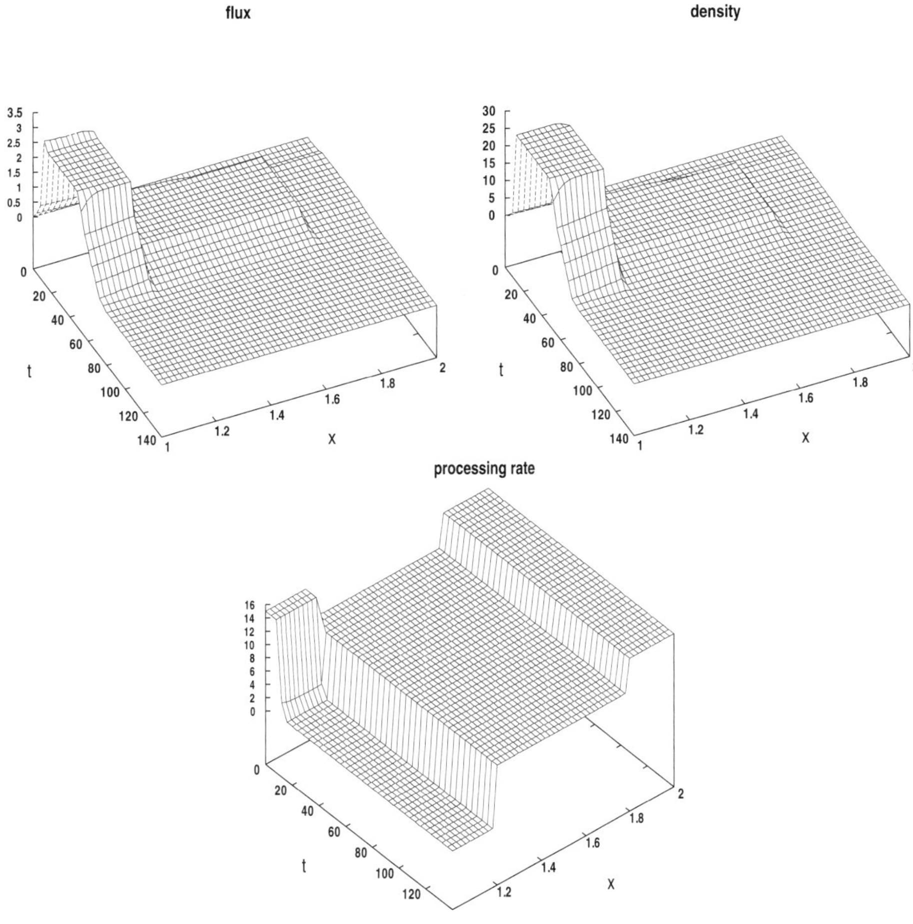

Figure 8.13. *Test* 1*: evolution on processors* 2, 3, 4 *of* f *(top left),* ρ *(top right), and* μ *(bottom) using SC*1*, with data in Table* 8.4 *and* $\varepsilon = 0.1$.

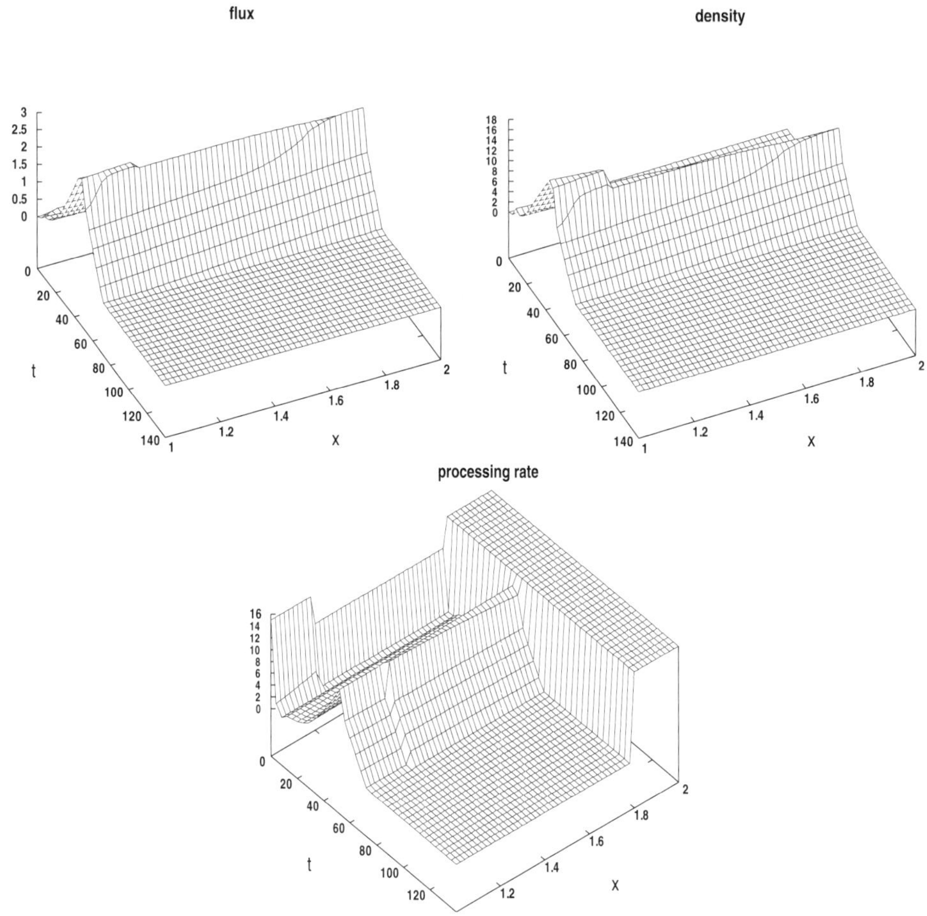

Figure 8.14. *Test* 1*: evolution on processors* 2, 3, 4 *of* f *(top left),* ρ *(top right), and* μ *(bottom) using SC*2*, with data in Table* 8.4 *and* $\varepsilon = 0.1$.

In Figures 8.15, 8.16, and 8.17 we depict the evolution in time on processors 2, 3, 4 of the flux, density, and processing rate, as obtained by the Riemann Solver SC3 with, respectively, $\varepsilon = 0.1$, $\varepsilon = 0.5$, and $\varepsilon = 0.01$. As shown by the graphics obtained, ε varying determines a different evolution. In particular, for ε tending to zero, the maximum values assumed by the flux and the density decrease.

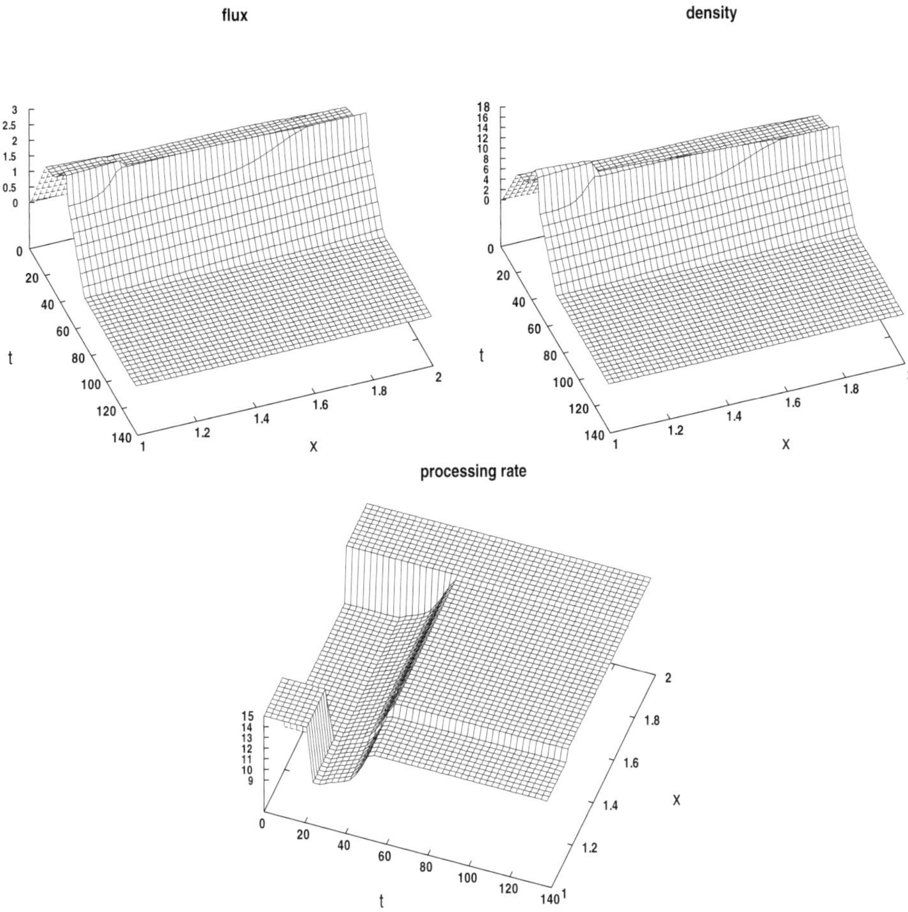

Figure 8.15. *Test* 1*: evolution on processors* 2, 3, 4 *of* f *(top left),* ρ *(top right), and* μ *(bottom) using SC3, with data in Table* 8.4 *and* $\varepsilon = 0.1$.

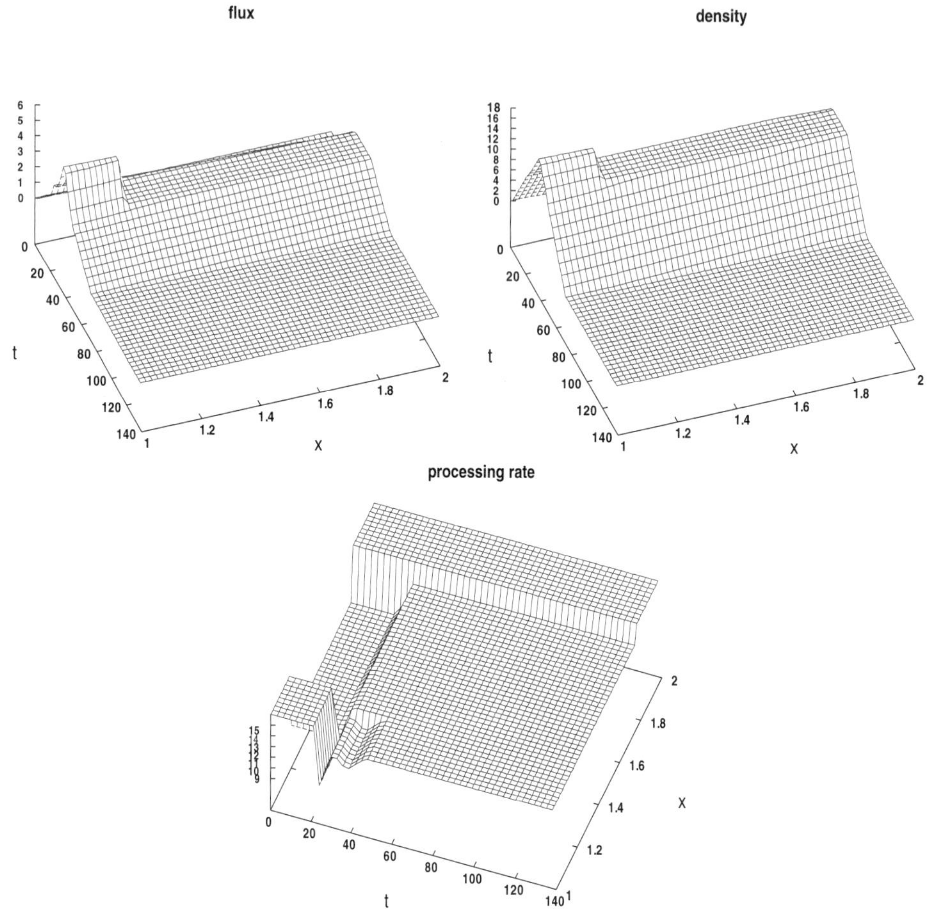

Figure 8.16. *Test* 1*: evolution on processors* 2, 3, 4 *of* f *(top left),* ρ *(top right), and* μ *(bottom) using SC3, with data in Table* 8.4 *and* $\varepsilon = 0.5$.

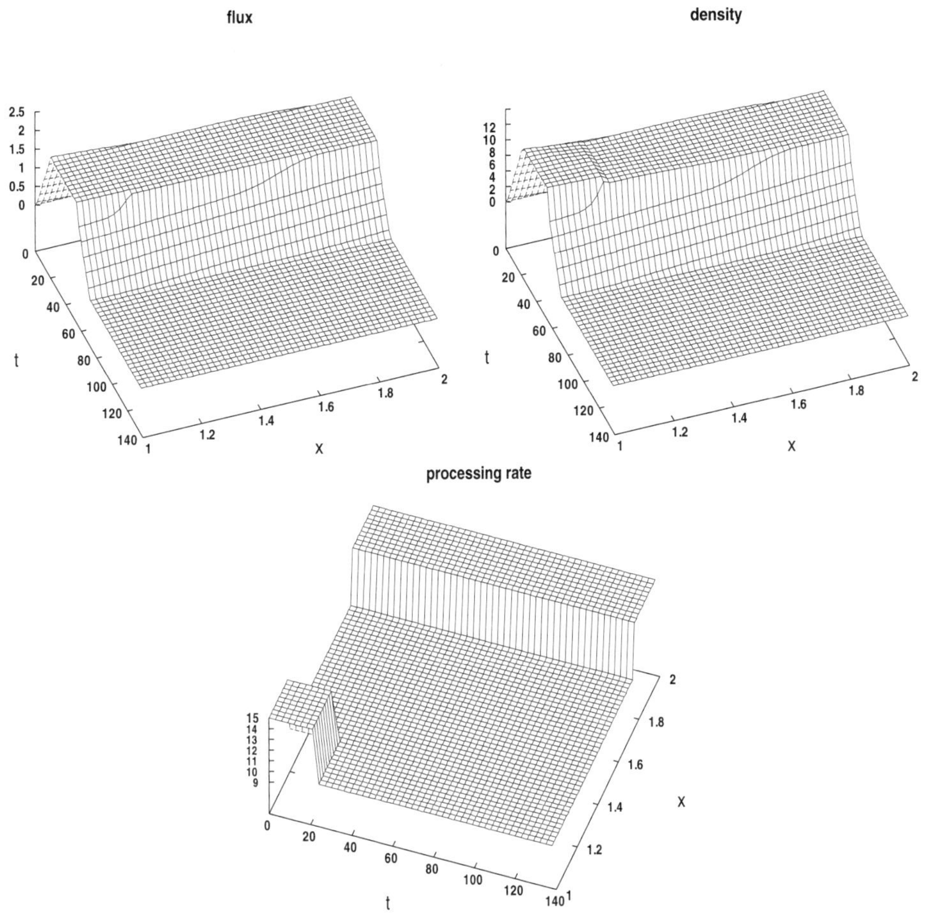

Figure 8.17. *Test* 1*: evolution on processors* 2, 3, 4 *of* f *(top left),* ρ *(top right), and* μ *(bottom) using SC3, with data in Table* 8.4 *and* $\varepsilon = 0.01$.

From the analysis of the graphics in Figures 8.15, 8.16, and 8.17, obtained by applying Riemann Solver SC3, we can deduce that adjustments of the processing rate determine the expected behavior of the density, also in accordance with results reported in [38].

Test 2. Referring to [37], we consider again a supply chain of $N=4$ arcs and impose the following initial and boundary data:

$$\rho_1(0,x)=\rho_2(0,x)=\rho_3(0,x)=\rho_4(0,x)=0,$$

$$\rho_1(t,0)=\frac{\mu_2}{2}\left(1+\sin\left(\frac{3\pi t}{T_{\max}}\right)\right),$$

where the space interval is $[0,6]$ and the observation time is $T_{\max}=20$, with $\Delta x=0.1$ and $\Delta t=0.05$. On each processor $e=1,2,3,4$ we assume $\mu(0,x)=\mu^e$ and incoming and outgoing boundary data are given by μ^e. Observe that even in this case the inflow profile $\rho_1(t,0)$ exceeds the maximum capacity of the processors.

Referring to [37], we make simulations setting parameters as in Table 8.5 and we assume to have default processing velocities on each processor, namely $m^e=1$, $e=1,2,3,4$.

Table 8.5. *Parameters of Test* 2.

Processor e	μ^e	L^e
1	99	1
2	15	1
3	10	3
4	8	1

In Figure 8.18 we depict the evolution in time of the flux, the density, and processing rate obtained by the Riemann Solver SC1 for $\varepsilon=0.1$, while in Figures 8.19 and 8.20 we show the behavior of the flux, density, and processing rate obtained, respectively, by the Riemann Solvers SC2 and SC3.

Let us make a comparison between the graphics in Figures 8.18 and 8.19. We observe that with solver SC2 the productivity collapses, thus provoking a lowering in the values of the flux and the density. On the other hand, SC1 maintains the level of productivity. Using solver SC3, which maximizes the flux and adjusts the processing rate if necessary, results are in accordance with those obtained in [37]; see Figure 8.20.

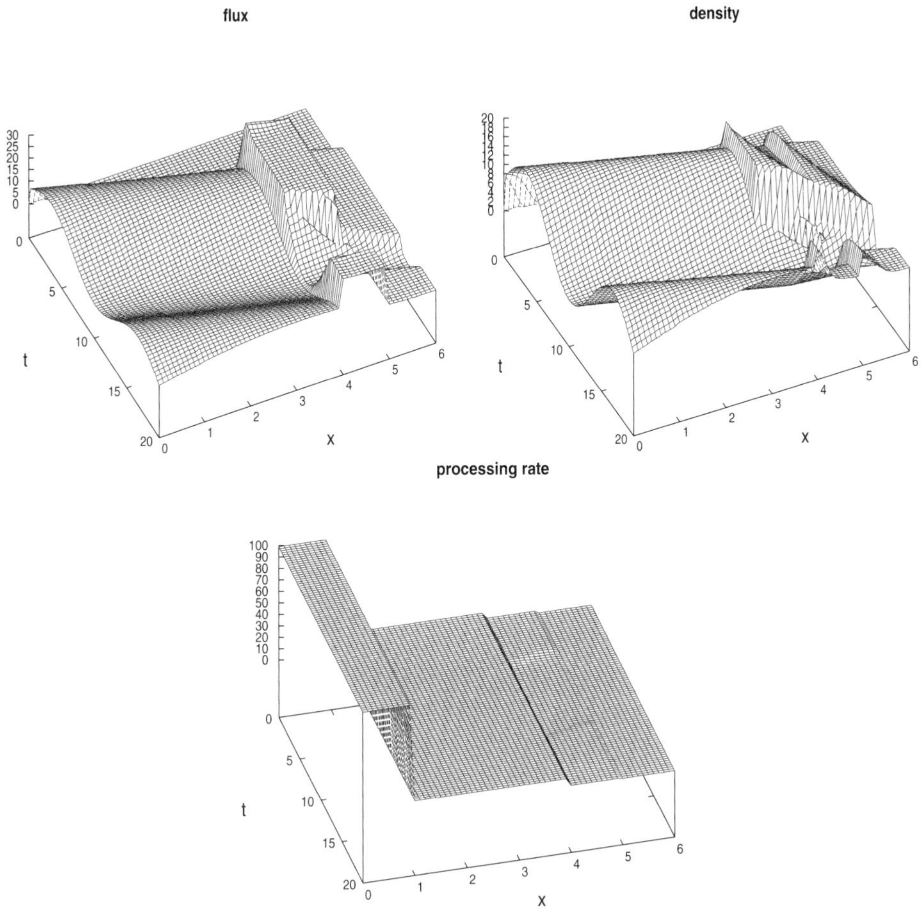

Figure 8.18. *Test* 2*: evolution of* f *(top left),* ρ *(top right), and* μ *(bottom) for the default velocities using SC*1*, with data in Table* 8.5 *and* $\varepsilon = 0.1$.

CPU Time

Now we are interested in the analysis of the CPU time. In particular, we want to compare the performances of programs based on the classical Godunov scheme (G) and on the fast Godunov scheme (FG) introduced in Section 8.1.4. To this aim, we report in Tables 8.6 and 8.7 the time of execution, expressed in seconds (s), of the simulation algorithm using SC3 and applied, respectively, to Test 1 and Test 2, when Δx (and consequently $\Delta t = \Delta x/2$) decreases.

In order to show the complexity, depending on the number of nodes as well as the space-time discretizations, of the simulation algorithm characterized by the proposed Riemann Solvers, we consider networks composed of a large number of nodes. In Table 8.8 we

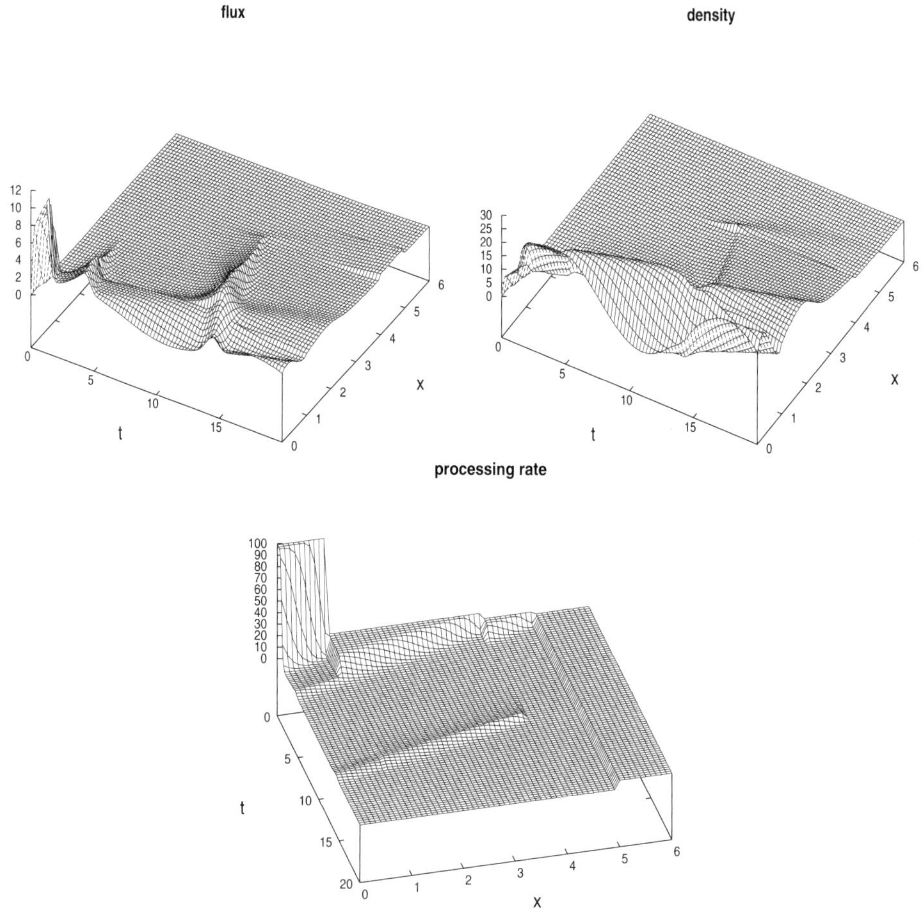

Figure 8.19. *Test* 2*: evolution of f (top left), ρ (top right), and μ (bottom) using SC*2*, with data in Table* 8.5 *and* $\varepsilon = 0.1$.

display the CPU time expressed in seconds of the simulation algorithm using SC3 applied to supply chains of $N = 100, 1000, 10000$ arcs.

From the analysis of the previous tables we can observe that the fast Godunov scheme (FG) allows us to save more than 40% of CPU time with respect to the classical Godunov scheme (G). Therefore we can conclude that FG performs better than G. We point out that the growth in the CPU time due to the increasing dimension of supply chain seems to be quadratic.

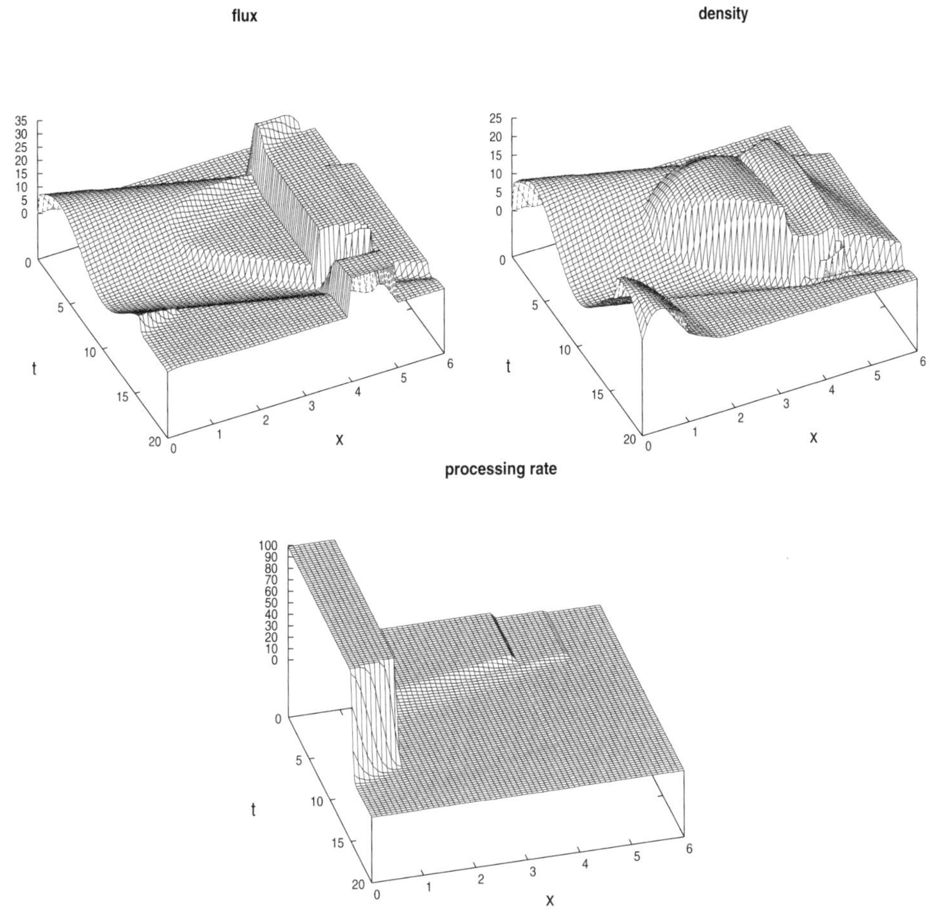

Figure 8.20. *Test 2: evolution of f (top left), ρ (top right), and μ (bottom) using SC3, with data in Table 8.5 and $\varepsilon = 0.1$.*

8.2 Optimization

As already mentioned in Chapter 7, optimization problems are an important aspect in managing supply chains, e.g., maximizing the output of a production process or minimizing used buffers. These optimization problems are originally formulated on a continuous level with constraints consisting of partial or ordinary differential equations. Usually an adjoint calculus is used for the efficient computation of the optimal control where the continuous optimality system is discretized and solved by a steepest descent method; see Section 7.3.2. Mathematically, we know that the negative gradient is a descent direction which is the steepest descent direction as well. The optimal solution can be computed since we know

Table 8.6. *CPU time for the schemes G and FG applied to Test 1; $T = 140$.*

CPU time		
	$T = 140$	
Δx	G	FG
0.1	0.06 s	0.04 s
0.05	0.19 s	0.12 s
0.025	0.71 s	0.4 s
0.0125	2.81 s	1.57 s
0.00625	13.5 s	7.0 s

Table 8.7. *CPU time for the schemes G and FG applied to Test 2; $T = 20$.*

CPU time		
	$T = 20$	
Δx	G	FG
0.2	0.01 s	0.01 s
0.1	0.02 s	0.02 s
0.05	0.07 s	0.04 s
0.025	0.25 s	0.16 s
0.0125	1.0 s	0.6 s

how to evaluate the gradient expression

$$D_{u^e} g(u) = K^e_{u^e}(q,u,t) + \sum_{e\in\mathcal{A}} [\phi^e - \psi^e(t,0)]\partial_{u^e} f^e(q^e) - \sum_{e\in\mathcal{A}} \sum_{\tilde{e}\in\delta_v^+} \phi^e u^{\tilde{e}} V^{\tilde{e}} \rho^{\tilde{e}}(t,1). \quad (8.10)$$

In the case of violating the box-constrained controls u_e during the performance of the algorithm, we have to deal with a simple projection to map the controls again to the admissible set U_{ad}.

1. Choose an initial control $u_e \in U_{ad}$.
2. Compute the solution of the state equation (7.3c)–(7.3d) and the adjoint equation (7.10)–(7.11).
3. Compute the gradient (8.10). If $P(D_{u^e} g(u)) = 0$, then STOP.
4. Choose the step size κ and update the control

$$\bar{u}_e := P(u_e - \kappa\, D_{u^e} g(u)) \in U_{ad}.$$

5. Set $u^e := \bar{u}^e$ and go to 2.

To ensure convergence of the descent method we can either use a fixed step size or employ a step size rule, e.g., the Armijo–Goldstein rule. See [56, 71] for a general overview.

Table 8.8. *CPU time of the algorithm for the schemes G and FG for* $T = 140$ *and* $N = 100, 1000, 10000$.

CPU time						
$T = 140$						
	$N = 100$		$N = 1000$		$N = 10000$	
Δx	G	FG	G	FG	G	FG
0.1	3.4 s	2.9 s	314.2 s	272.6 s	30981.9 s	26816.2 s
0.05	7.5 s	6.8 s	603.2 s	525.5 s	59629.1 s	51088 s
0.025	19 s	16.2 s	1422.9 s	1208.7 s	138310.7 s	117502.4 s
0.0125	57.2 s	44.1 s	3558 s	2900.6 s	365281.5 s	282005.8 s
0.00625	202.3 s	138.9 s	10010.9 s	7832.2 s	1059316.3 s	761415.7 s

Moreover, we know that for our supply network model (7.3) the discretization can be chosen such that the optimization problem is in fact a mixed-integer programming problem. This is mainly due to the fact that the governing dynamics in the supply network are linear in the state (but not in the control) variables. The proposed MIP model is stated in (7.40). The essential difference to the adjoint approach is the reformulation of the constraint

$$f^e(q^e(t)) = \min\left\{\mu^e; \frac{q^e(t)}{\epsilon}\right\}$$

by introducing binary variables ζ_j^e. This finally leads to an MIP and not just a linear programming model. For solving the MIP model the standard optimization solver ILOG CPLEX [55] with default settings is used. A common algorithm to find optimal solutions is the so-called Branch-and-Bound algorithm. The method is based on a systematic generation of subproblems (Branching) which are analyzed by computing upper and lower bounds of the respective cost functional values (Bounding). In order to find a lower bound for the original problem, the LP relaxation of (7.40) is solved; i.e., the binary variables are assumed to be real variables such as in an ordinary LP. If the solution of the relaxed problem does not fulfill all integrality constraints of the original problem, it is split into two subproblems by fixing one of the binary variables to 1 and 0, respectively. In general, appropriate heuristical algorithms can be used to provide an upper bound for the root problem that is the cost functional value of a feasible solution. Repeating this procedure iteratively, we end up with a Branch-and-Bound tree whose nodes present the subproblems; cf. Figure 8.21.

In most cases, a subproblem can be pruned, which means that no further branching is necessary at this point. We distinguish between three kinds of pruning:

1. Pruning by optimality: If the optimal solution of a subproblem has been found, no further branching on that node is necessary.
2. Pruning by bound: If the lower bound of the optimal objective function value of a subproblem is greater than a global upper bound being defined as the minimum of all upper bounds that have been found so far, the optimal solution is not included in this subproblem anymore.
3. Pruning by infeasibility: A subproblem does not contain any feasible solution.

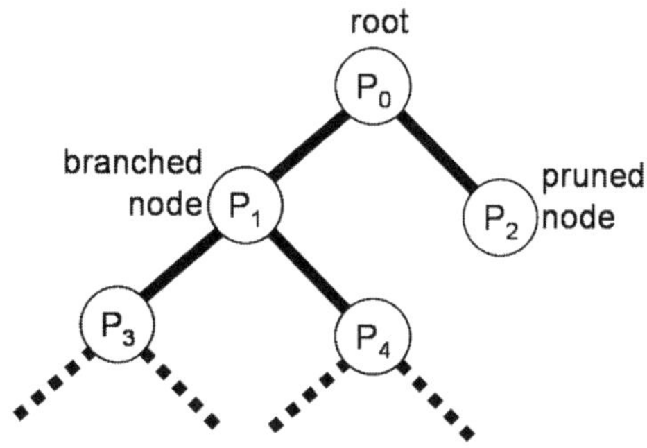

Figure 8.21. *Branch-and-Bound tree where P_i denotes a single subproblem.*

The faster the Branch-and-Bound algorithm terminates, the sharper are the bounds; i.e., nodes can be pruned and the size of the tree is kept small.

In the next sections, we present computational results of the adjoint-based method compared with the discrete approaches (MIP and LP). Furthermore, we report on properties on the mixed-integer model and show the efficiency of the LP.

8.2.1 Qualitative Comparison of Approaches

At first, we compare the qualitative behavior of results computed by the adjoint approach and the mixed-integer model. From above we know when considering optimization problems such as (7.3) that we have the possibility to either apply the mixed-integer approach leading to (7.40) or to solve the continuous optimality system (7.3). We emphasize that if we use the same discretization as proposed in Chapter 7 for the continuous problem, we obtain exactly the same optimality system. The difference between the adjoint approach and the MIP (7.40) is just the solution method.

For our numerical comparison, we stick to the following real-world example. We use the network given in Figure 8.22 and its abstract form in Figure 8.23. The network consists of 12 processors and queues, and we have the six free controls $A^{1,3}(t)$–$A^{1,7}(t)$ and $A^{2,10}(t)$. The first arc in Figure 8.23 is artificial and is used to prescribe the inflow profile only. In Figure 8.22, a layout of a supply chain producing toothbrushes is shown.

We are interested in optimizing the number of pallets on which the half-finished toothbrushes are carried through the network. In detail, pallets are fed into the network at processor 2. There, each pallet is assembled before it is processed in processors 3–11. The finished toothbrushes are removed in processor 12, and the pallets start circling again. We assume the number of pallets is given by an inflow profile which we prescribe on arc 1. Here, the inflow data is given by a step function: $f^{in}(t) := 0.852$ for $0 \leq t \leq N_T/2$ and $f^{in}(t) := 0$ otherwise. Further data is given in Table 8.9.

The aim of the optimization problem is maximizing the outflow of processor 12, i.e., $f_j^{12,b}$. This means optimizing the number of pallets (in particular finished toothbrushes) passed to processor 12 under the constraints introduced in Section 7.4.1. The mixed-integer

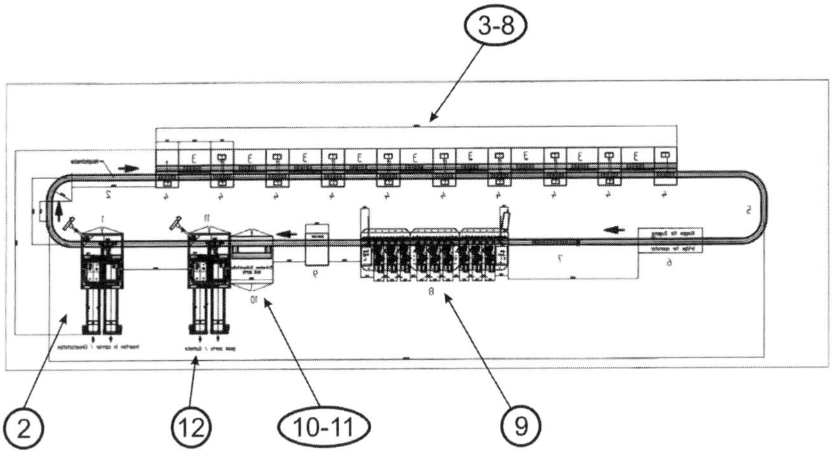

Figure 8.22. *Engineering layout of a production line.*

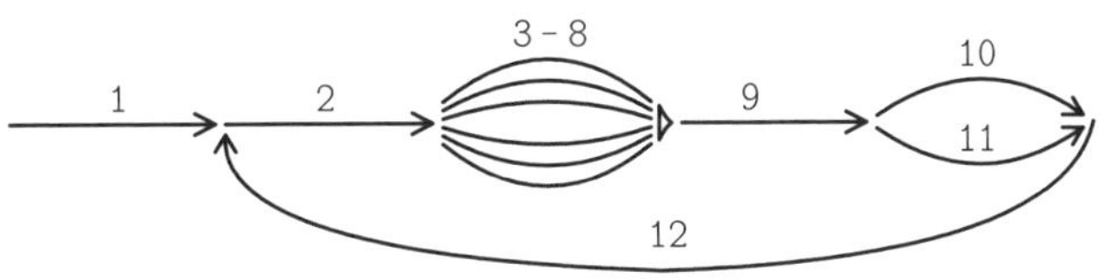

Figure 8.23. *Corresponding mathematical network.*

problem (7.39) is adapted to the objective function

$$\sum_{j=1}^{N_T} -\frac{1}{j+1} \mathrm{f}_j^{12,b}. \tag{8.11}$$

For all computations we fix the smoothing parameter $\epsilon = 1$, the time horizon $T = 200$, and the constant $\mathcal{M}$.

In Figures 8.24, 8.25, and 8.26 we present results for the optimal routing problem by pointing out similarities and differences between the adjoint and discrete approaches. The computation of the adjoint approach takes 11.15 seconds using 18 iterations and for the MIP 11.16 seconds using 551 iterations. In the adjoint approach, we terminate the iteration if the relative error of two consecutive iterates is less than $tol := 1e^{-6}$—consistent with the default accuracy in ILOG CPLEX [55]. Both approaches yield an optimal functional value of $g^*(A^{v,e}) = -0.19$. In Figure 8.24 we plot the optimal outflow profile computed by the two approaches. We observe that for this particular example the curves coincide.

However, the computed optimal controls, and consequently the time evolution of the queues, differ considerably. In Figures 8.25 and 8.26 we plot the optimal controls feeding parts into processors 3–10 for the MIP and the adjoint approach, respectively. Since the incoming flux is higher than the sum of all production rates on these arcs, queues will grow and all processors are filled. We further observe that, although the production rates μ^3–μ^8

Table 8.9. *Relevant data split according to maximal processing rates μ^e, lengths L^e, and processing velocities V^e.*

Processor	μ^e	V^e	L^e
1	100	0.01333	1
2	0.71	0.35714	1.5
3–8	0.06666	0.01333	1
9	0.71	0.04762	3
10–11	0.24	0.119	1.5
12	0.71	0.35714	1.5

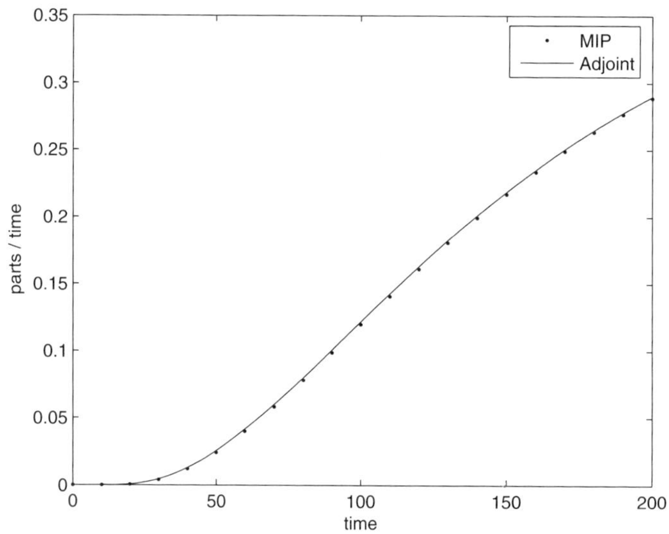

Figure 8.24. *Optimal output for processor* 12.

have the same value, the flux is not equally distributed. In the case of the adjoint approach the distribution rates $A^{1,3}$–$A^{1,7}$, according to the degree of freedom, are equal and the last one is computed using Definition 5.2. There is no similar pattern recognizable in the MIP approach.

Although the optimal functional values coincide we see that we do not have a unique minimizer to our optimal control problem.

8.2.2 Convergence Results for the MIP

In this section, we investigate the behavior of the solution for finer space discretizations. Therefore, we introduce an equidistant grid in space. This means each arc is now discretized

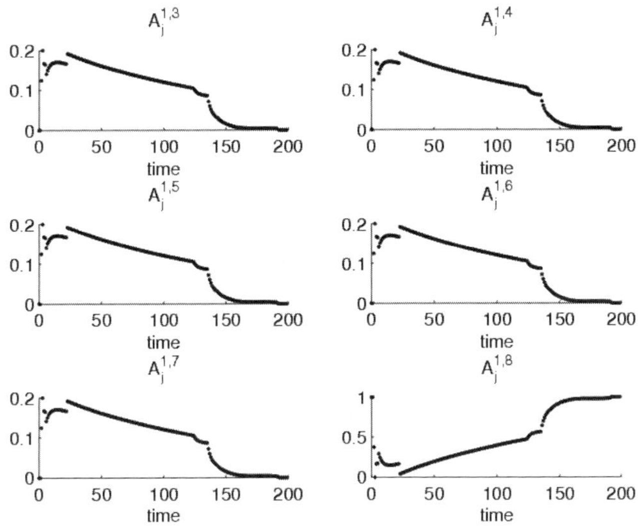

Figure 8.25. *Plot of the distribution rate* $A_j^{1,e}$, $e = 3,\ldots,8$, *computed by the adjoint approach.*

with D internal points such that $x_i \in [a^e, b^e]$ represents one space discretization point and $\Delta x = L^e/(D-1)$ the spatial grid size. We denote by $\mathrm{f}_j^{e,i}$, $j = 1,\ldots,D$, the flux inside the processor measured at points $x_i \in [a^e, b^e]$ and by $\mathrm{f}_j^{e,1} = \mathrm{f}_j^{e,a}$ the incoming flux into the processor at $x = a^e$. All fluxes $\mathrm{f}_j^{e,i}$ are bounded by the maximal capacity on this arc:

$$0 \le \mathrm{f}_j^{e,i} \le \mu^e \quad \forall e, \forall i, \forall j. \tag{8.12}$$

Obviously, $D = 2$ yields the two-point upwind discretization introduced in Section 7.4 with $\Delta x = L^e$.

Using the new notation, (7.31) can be rewritten as

$$\mathrm{f}_{j+1}^{e,i} = \mathrm{f}_j^{e,i} + \frac{\Delta t}{\Delta x} V^e \left(\mathrm{f}_j^{e,i-1} - \mathrm{f}_j^{e,i} \right) \quad \forall e, j, \quad i = 2,\ldots,D. \tag{8.13}$$

Furthermore, condition (7.37) changes to

$$\Delta t = \min\{\Delta x / V^e : e \in \mathcal{A}\}, \tag{8.14}$$

which induces smaller time intervals and, consequently, larger systems. For the computation, all other network equations remain unchanged.

The underlying network for our convergence analysis is depicted in Figure 8.27. We use the parameter setting $V^e = 1$, $L^e = 1$ for all $e \in \mathcal{A}$ and $\mu = (100, 40, 30, 20, 20, 5, 10, 10)$, where $\mu = (\mu^1, \ldots, \mu^8)$.

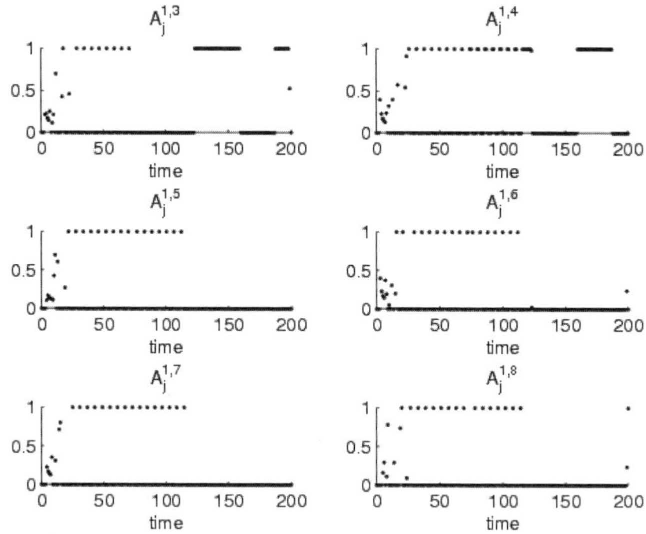

Figure 8.26. *Plot of the distribution rate* $A_j^{1,e}$, $e = 3, \ldots, 8$, *computed by the MIP.*

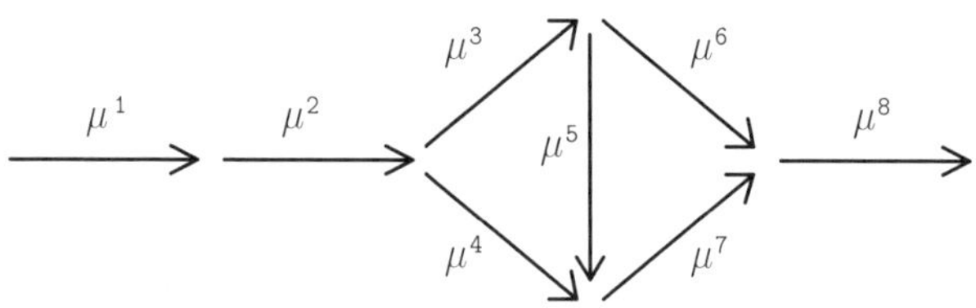

Figure 8.27. *Maximal processing capacities on each arc.*

We consider the following optimization problem with $\mathrm{f}_j^{e,D} = \mathrm{f}_j^{e,b}$:

$$\sum_{j=1}^{N_T} -\frac{1}{j+1} \mathrm{f}_j^{8,D}, \tag{8.15}$$

$$\begin{array}{ll} \min\limits_{A_t^{v,e}, v \in \mathcal{V}_d} & (8.15) \\ \text{subject to} & (8.13), (7.32), (7.33), (8.12). \end{array} \tag{8.16}$$

The inflow into the network is always prescribed at the first arc. We choose the continuous inflow profile as shown in Figure 8.28.

Obviously, it holds that $\int_0^T f^{in}(t)\,dt = 450$. The conservation of mass guarantees that the flux given into the network also leaves the network after several time steps (for T large enough); cf. Figure 8.28. Therefore, this value can be used as a reference value for the comparison of different space discretizations.

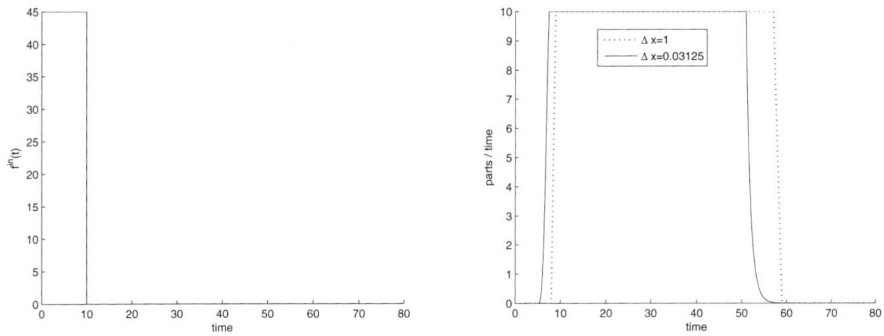

Figure 8.28. *Inflow $f^{in}(t)$ into the system (left). Outflow $\mathrm{f}_j^{8,D}$ of the system plotted for different space discretizations $\Delta x = 1$ and $\Delta x = 0.03125$ (right).*

Table 8.10. *Convergence results for finer space discretizations.*

Δx	$\Delta t \sum_j \mathrm{f}_j^{8,D}$
1	495
0.5	472.5
0.25	461.25
0.125	455.625
0.0625	452.8125
0.03125	451.4063

In Table 8.10, we observe that for smaller step sizes Δx the values of $\Delta t \sum_j \mathrm{f}_j^{8,D}$ converge to the value of the integral. We also recognize that the difference between two computed values decreases linearly with the space grid size.

Another example using the same setting where only the length of one arc is changed, $L^7 = 10$, shows that the solution of the coarse discretization $\Delta x = 1$ is smoothed out on this arc, whereas the solution of the fine discretization $\Delta x = 0.03125$ converges to the limit given by $\mu^7 = 10$; see Figure 8.29.

This result also emphasizes the use of mixed-integer models for solving partial differential equation constrained optimal control problems.

8.2.3 Computational Times

Real-World Network

We proceed with a comparison of computational times of the adjoint-based approach, the mixed-integer formulation, and the LP. Our computations are performed on the network of Figure 8.23 with default parameters $V^\varepsilon = L^\varepsilon = 1$, $\varepsilon = 2, \ldots, 12$, $\epsilon = 1$ and time horizon $T = 200$. We increase the number of time intervals N_T and solve both the MIP and the LP by using the commercial solver ILOG CPLEX [55].

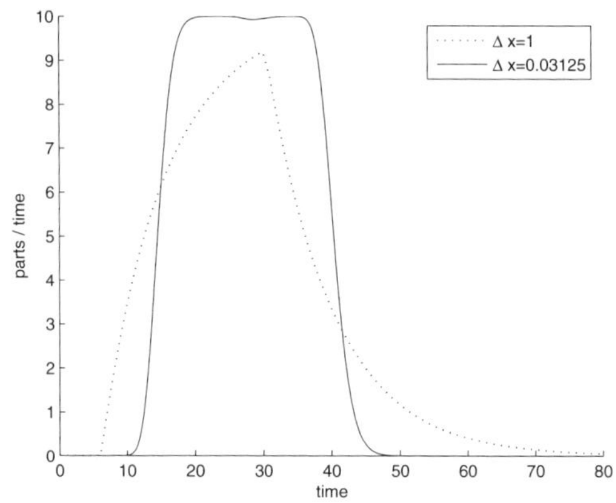

Figure 8.29. *Evolution of the flux* $f_j^{7,D}$ *plotted for different space discretizations* $\Delta x = 1$ *and* $\Delta x = 0.03125$.

Table 8.11. *CPU times in seconds for the real-world network in Figure* 8.23.

N_T	Adjoint	MIP	LP
200	7.31	5.52	0.58
400	26.10	17.06	2.17
800	45.10	68.09	6.56
2000	124.58	592.61	41.41
4000	333.13	infeasible	214.45

Obviously, the most efficient solving technique for the optimal control problem stated in Chapter 7 is the LP which is even solvable for large instances. At present, the MIP fails to compute a solution for $\Delta t \leq 0.05$ since the system becomes too large and the preprocessing procedure produces infeasible solutions. As Table 8.11 further indicates, the MIP is faster than the adjoint calculus if up to approximately 600 time steps are used (corresponding to $\Delta t \in [0.\overline{3}, 1]$). For small instances, i.e., 30 000 real variables, the LP is solved within a few seconds. As N_T increases, the adjoint approach becomes more attractive, also compared to the LP. For values of $\Delta t < 0.\overline{3}$ it computes an optimal solution faster than the MIP.

Block Networks

As a further example, we solve the MIP and the LP on connected and large-scale networks to gain further insights into the computational complexity of both approaches. The network we use is a composition of standard building blocks. Each block has several inflow and outflow arcs such that the flow has to be redistributed at each vertex. We consider networks of $k \times 3$ interior network vertices, where k runs from 1 to 8. An illustration of such a block network is given in Figure 8.30.

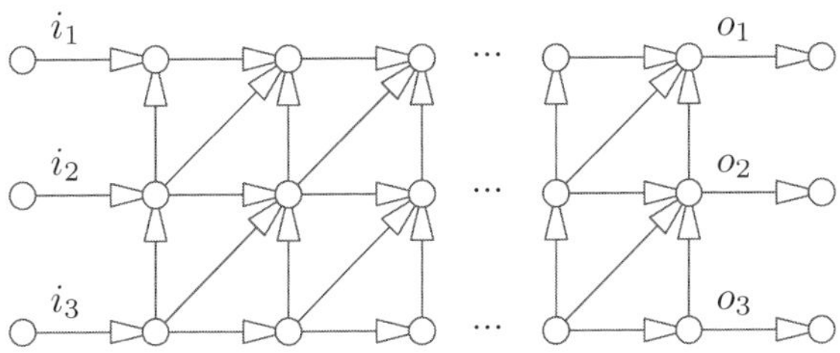

Figure 8.30. *$k \times 3$ block network.*

The values L^e, V^e, μ^e are set to 1 for all suppliers except the inflow arcs $i_1, \ldots, i_3$. Here, we choose maximal processing rates of $\mu^e = 100$. The total number of time steps is $N_T = 800$. Inflow profiles (box functions) are prescribed on all ingoing arcs $i_1, \ldots, i_3$. For different sizes of networks, $k \in \{1, \ldots, 8\}$, we evaluate the presolve and the total solution time for the corresponding MIP and LP models. The results are given in Table 8.12. Clearly, the larger the network, the more time that is needed to compute a globally optimal solution. Further, this test shows the behavior of solution times for large problem instances. It can be seen that both times (presolve and solution time) consumed by CPLEX grow with the network size k. The MIP produces infeasible results for $k > 7$.

Table 8.12. *Presolve and solution times for the MIP and the LP in seconds for $k \times 3$ block networks.*

k	Presolve MIP	Solution MIP	Presolve LP	Solution LP
1	78.87	105.33	0.11	6.26
2	159.09	201.64	0.32	33.54
3	236.59	332.78	0.54	76.91
4	352.08	555.11	0.75	127.24
5	477.50	810.44	0.99	232.68
6	590.09	1163.25	1.22	348.53
7	983.55	1891.90	1.42	502.72
8	907.01	infeasible	1.73	768.47

Figure 8.31 shows that the solution times computed by the MIP grow exponentially with the size of the instances. However, the LP produces solution times which grow nearly linearly.

8.3 Application

Finally, we conclude the chapter on numerical results with a more complex flow network. This example consists of 418 arcs and 233 vertices. The network is highly interconnected and has four entries and four exits. The graph of the network is depicted in Figure 8.32. We consider a single product flow and prescribe material flows on the entries. In view

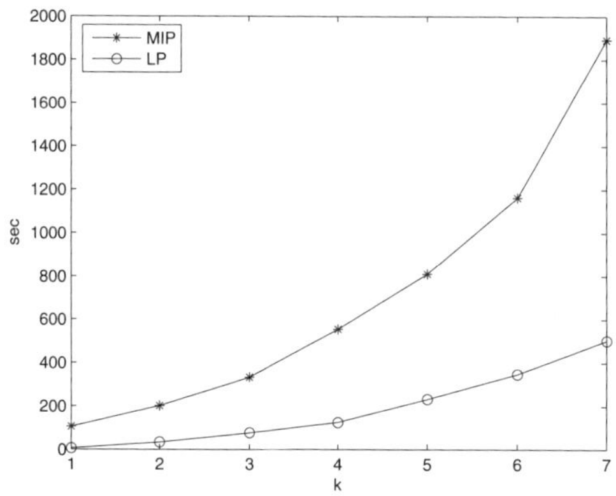

Figure 8.31. *Visualization of the computing times in Table* 8.12.

of the optimization we ask for an optimal load of the network over time as well as for a prediction of the time evolution of the network. For simplicity we assume unlimited buffers. To compare optimal and nonoptimized flows in the network we depicted the actual flow on selected arcs of the network. The amount of flow is indicated by bold arc lines. The arcs selected belong to the box in Figure 8.32. To show the dynamics of the network flows we present results at selected times. A simulation result is depicted in Figure 8.33. In this case the distribution of goods among suppliers is randomly chosen. The load of the arcs and the time evolution has to be compared with the optimal load of the network. The latter is obtained by solving the optimization problem (7.40) presented in the previous section. The time evolution of the corresponding flows is depicted in Figure 8.34. In both cases the supply network has to process the same number of goods. When distributing with the optimal controls, the parts are moved faster towards possible exits. Therefore, the total load over time is expected to be significantly smaller compared to the nonoptimal case. This can be observed in the results. Using the optimal distribution rates the load of the network in the selected area is significantly reduced: For example at time $j = 50$ only a few arcs are currently loaded with parts compared to the fully loaded network in Figure 8.33. This example fails to be computable using the MIP model (7.40) and a time limit of one day. However, the corresponding LP is solved in a few hours. The results shows that in fact complex networks with the continuous dynamics can be optimized and time-optimal distribution rates can be obtained.

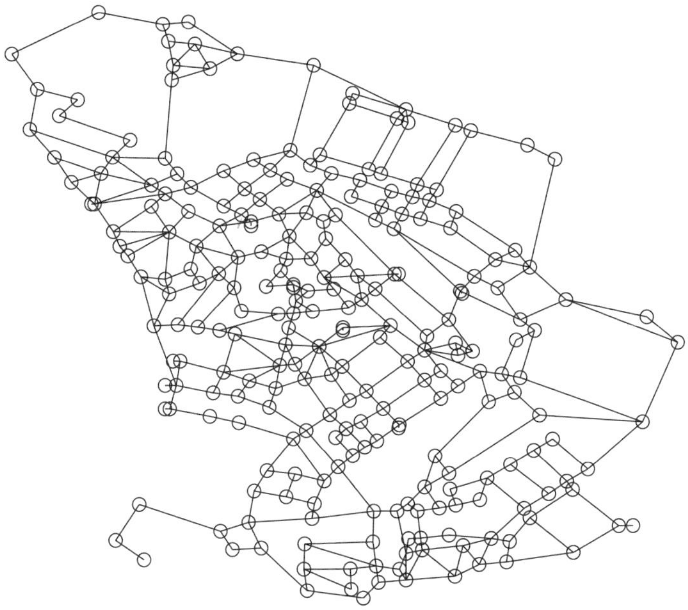

Figure 8.32. *A supply network consisting of* 418 *arcs and* 233 *vertices. The box indicates the clipping area for simulation and optimization results.*

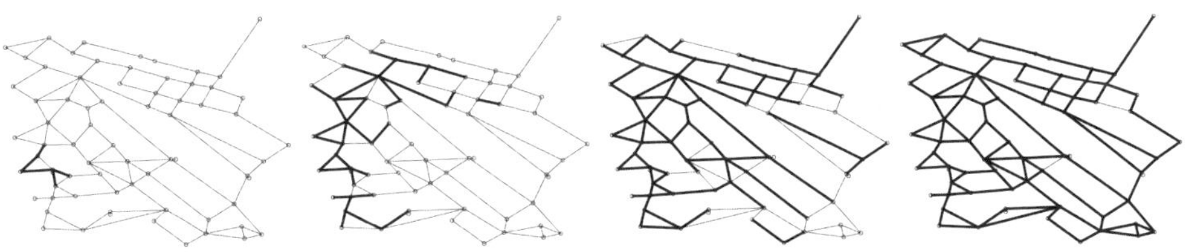

Figure 8.33. *Simulation results for* $j = 10, 23, 36, 50$ *(from left to right).*

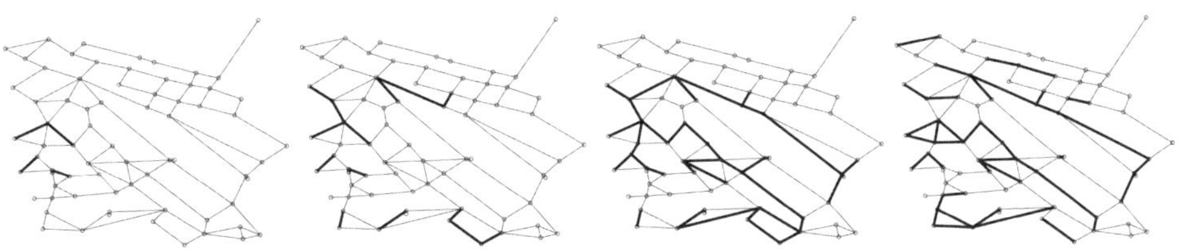

Figure 8.34. *Optimization results for* $j = 10, 23, 36, 50$ *(from left to right).*

Bibliography

[1] D. ARMBRUSTER, C. DE BEER, M. FREITAG, T. JAGALSKI, AND C. RINGHOFER, *Autonomous control of production networks using a pheromone approach*, Phys. A, 363 (2006), pp. 104–114.

[2] D. ARMBRUSTER, P. DEGOND, AND C. RINGHOFER, *A model for the dynamics of large queuing networks and supply chains*, SIAM J. Appl. Math., 66 (2006), pp. 896–920.

[3] ——, *Kinetic and fluid models for supply chains supporting policy attributes*, Bull. Inst. Math. Acad. Sin. (N.S.), 2 (2007), pp. 433–460.

[4] D. ARMBRUSTER, D. MARTHALER, AND C. RINGHOFER, *Kinetic and fluid model hierarchies for supply chains*, Multiscale Model. Simul., 2 (2003), pp. 43–61.

[5] A. AW AND M. RASCLE, *Resurrection of "second order" models of traffic flow*, SIAM J. Appl. Math., 60 (2000), pp. 916–938.

[6] M. K. BANDA, M. HERTY, AND A. KLAR, *Coupling conditions for gas networks governed by the isothermal Euler equations*, Netw. Heterog. Media, 1 (2006), pp. 295–314.

[7] J. BANKS AND J. S. CARSON, *Discrete-Event System Simulation*, Prentice–Hall International Series in Industrial and Systems Engineering, Prentice–Hall, Englewood Cliffs, NJ, 1984.

[8] W.-J. BAUMOL, *Economic Dynamics*, Macmillan, New York, 1970.

[9] G. BOLCH, S. GREINER, H. DE MEER, AND K. S. TRIVEDI, *Queueing Networks and Markov Chains. Modeling and Performance Evaluation with Computer Science Applications*, 2nd ed., John Wiley & Sons, Hoboken, NJ, 2006.

[10] A. BRESSAN, *Hyperbolic Systems of Conservation Laws*, Oxford University Press, Oxford, UK, 2000.

[11] A. BRESSAN, G. CRASTA, AND B. PICCOLI, *Well-posedness of the Cauchy problem for $n \times n$ systems of conservation laws*, Mem. Amer. Math. Soc., 146 (2000), no. 694.

[12] A. BRESSAN, T.-P. LIU, AND T. YANG, *L^1 stability estimates for $n \times n$ conservation laws*, Arch. Ration. Mech. Anal., 149 (1999), pp. 1–22.

[13] G. BRETTI, C. D'APICE, R. MANZO, AND B. PICCOLI, *A continuum-discrete model for supply chains dynamics*, Netw. Heterog. Media, 2 (2007), pp. 661–694.

[14] R. Burkhard, M. Dell'Amico, and S. Martello, *Assignment Problems*, SIAM, Philadelphia, 2009.

[15] C. Cercignani, R. Illner, and M. Pulvirenti, *The Mathematical Theory of Dilute Gases*, Applied Mathematical Sciences 106, Springer-Verlag, New York, 1994.

[16] H. Chen and D. D. Yao, *Fundamentals of Queueing Networks. Performance, Asymptotics, and Optimization*, Applications of Mathematics 46, Springer, New York, 2001.

[17] G. M. Coclite, M. Garavello, and B. Piccoli, *Traffic flow on a road network*, SIAM J. Math. Anal., 36 (2005), pp. 1862–1886.

[18] R. M. Colombo and M. Garavello, *A well posed Riemann problem for the p-system at a junction*, Netw. Heterog. Media, 1 (2006), pp. 495–511.

[19] R. M. Colombo, M. Herty, and V. Sachers, *On 2×2 conservation laws at a junction*, SIAM J. Math. Anal., 40 (2008), pp. 605–622.

[20] C. M. Dafermos, *Polygonal approximations of solutions of the initial value problem for a conservation law*, J. Math. Anal. Appl., 38 (1972), pp. 33–41.

[21] ——, *Hyperbolic Conservation Laws in Continuum Physics*, Springer, Berlin, Heidelberg, New York, 2000.

[22] C. F. Daganzo, *Requiem for second order fluid approximations of traffic flow*, Trans. Res. B, 29 (1995), pp. 277–286.

[23] ——, *A continuum theory of traffic dynamics for freeways with special lanes*, Trans. Res. B, 31 (1997), pp. 83–102.

[24] ——, *A Theory of Supply Chains*, Lecture Notes in Economics and Mathematical Systems 526, Springer, Berlin, 2003.

[25] C. D'Apice and R. Manzo, *Calculation of predicted average packet delay and its application for flow control in data network*, J. Inf. Optim. Sci., 27 (2006), pp. 411–423.

[26] ——, *A fluid dynamic model for supply chains*, Netw. Heterog. Media, 1 (2006), pp. 379–398.

[27] C. D'Apice, R. Manzo, and B. Piccoli, *Existence of solutions to Cauchy problems for a mixed continuum-discrete model for supply chains and networks*, J. Math. Anal. Appl., 362 (2010), pp. 374–386.

[28] ——, *Modelling supply networks with partial differential equations*, Quart. Appl. Math., 67 (2009), pp. 419–440.

[29] P. Degond, S. Göttlich, M. Herty, and A. Klar, *A network model for supply chains with multiple policies*, Multiscale Model. Simul., 6 (2007), pp. 820–837.

[30] J. W. Forrester, *Industrial Dynamics*, MIT Press, Cambridge, MA, 1964.

[31] A. FÜGENSCHUH, S. GÖTTLICH, AND M. HERTY, *A new modeling approach for an integrated simulation and optimization of production networks*, H.O. Günther, D. Mattfeld, L. Suhl (eds.), Supply Networks and Logistics Management, Physica-Verlag, Heidelberg, 2007, pp. 45–60.

[32] A. FÜGENSCHUH, M. HERTY, A. KLAR, AND A. MARTIN, *Combinatorial and continuous models for the optimization of traffic flows on networks*, SIAM J. Optim., 16 (2005), pp. 1155–1176.

[33] M. GARAVELLO AND B. PICCOLI, *Traffic Flow on Networks*, American Institute of Mathematical Sciences, Springfield, MO, 2006.

[34] J. GLIMM, *Solutions in the large for nonlinear hyperbolic systems of equations*, Comm. Pure Appl. Math., 18 (1965), pp. 697–715.

[35] E. GODLEWSKI AND P.-A. RAVIART, *Hyperbolic Systems of Conservation Laws*, Mathématiques & Applications [Mathematics and Applications] 3/4. Ellipses, Paris, 1991.

[36] ——, *Numerical Approximation of Hyperbolic Systems of Conservation Laws*, Springer, New York, 1996.

[37] S. GÖTTLICH, M. HERTY, AND A. KLAR, *Modelling and optimization of supply chains on complex networks*, Commun. Math. Sci., 4 (2006), pp. 315–330.

[38] ——, *Network models for supply chains*, Commun. Math. Sci., 3 (2005), pp. 545–559.

[39] D. HELBING, *Improved fluid dynamic model for vehicular traffic*, Phys. Rev. E, 51 (1995), p. 3164.

[40] ——, *Verkehrsdynamik*, Springer, Berlin, Heidelberg, New York, 1997.

[41] ——, *Production, supply, and traffic systems: A unified description*, S. P. Hoogendoorn et al. (eds.), Traffic and Granular Flow '03, Springer, Berlin, 2005, pp. 173–188.

[42] D. HELBING, D. ARMBRUSTER, A. MIKHAILOV, AND E. LEFEBER, *Information and material flows in complex networks*, Phys. A, 363 (2006), pp. xi–xvi.

[43] D. HELBING AND S. LÄMMER, *Supply and production networks: From the bullwhip effect to business cycles*, D. Armbruster, A. S. Mikhailov, and K. Kaneko (eds.), Networks of Interacting Machines: Production Organization in Complex Industrial Systems and Biological Cells, World Scientific, Singapore, 2005, pp. 33–66.

[44] D. HELBING, S. LÄMMER, T. SEIDEL, P. SEBA, AND T. PLATKOWSKI, *Physics, stability and dynamics of supply networks*, Phys. Rev. E, 70 (2004), 066116.

[45] M. HERTY, *Modeling, simulation and optimization of gas networks with compressors*, Netw. Heterog. Media, 2 (2007), pp. 81–97.

[46] M. HERTY, M. GUGAT, A. KLAR, AND G. LEUGERING, *Optimal control for traffic flow networks*, J. Optim. Theory Appl., 126 (2005), pp. 589–616.

[47] M. HERTY AND A. KLAR, *Modeling, simulation, and optimization of traffic flow networks*, SIAM J. Sci. Comput., 25 (2004), pp. 1066–1087.

[48] M. HERTY, S. MOUTARI, AND M. RASCLE, *Optimization criteria for modelling intersections of vehicular traffic flow*, Netw. Heterog. Media, 1 (2006), pp. 275–294.

[49] M. HERTY, R. PINNAU, AND G. THÖMMES, *Asymptotic and discrete concepts for optimal control in radiative transfer*, ZAMM Z. Angew Math. Mech., 87 (2007), pp. 333–347.

[50] M. HERTY AND C. RINGHOFER, *Optimization for supply chain models with policies*, Phys. A, 380 (2007), pp. 651–664.

[51] M. HINZE AND K. KUNISCH, *Second order methods for optimal control of time-dependent fluid flow*, SIAM J. Control Optim., 40 (2001), pp. 925–946.

[52] M. HINZE AND R. PINNAU, *An optimal control approach to semiconductor design*, Math. Models Methods Appl. Sci., 12 (2002), pp. 89–107.

[53] H. HOLDEN AND N. H. RISEBRO, *A mathematical model of traffic flow on a network of unidirectional roads*, SIAM J. Math. Anal., 26 (1995), pp. 999–1017.

[54] ——, *Front Tracking for Hyperbolic Conservation Laws*, Springer, New York, Berlin, Heidelberg, 2002.

[55] ILOG CPLEX DIVISION, *Using the CPLEX Callable Library*, 889 Alder Avenue, Suite 200, Incline Village, NV 89451, USA, 2000. Information available online from http://www.cplex.com.

[56] C. T. KELLEY, *Iterative Methods for Optimization*, Frontiers in Applied Mathematics 18, SIAM, Philadelphia, 1999.

[57] C. KIRCHNER, M. HERTY, S. GÖTTLICH, AND A. KLAR, *Optimal control for continuous supply network models*, Netw. Heterog. Media, 1 (2006), pp. 675–688.

[58] A. KLAR AND R. WEGENER, *Kinetic derivation of macroscopic anticipation models for vehicular traffic*, SIAM J. Appl. Math., 60 (2000), pp. 1749–1766.

[59] S. N. KRUZKOV, *First order quasilinear equations in several independent variables*, Math. USSR-Sb., 10 (1970), pp. 217–243.

[60] C. D. LAIRD, L. T. BIEGLER, B. G. VAN BLOEMEN WAANDERS, AND R. A. BARTLETT, *Contamination source determination for water networks.*, J. Water Resour. Plann. Manage., 131 (2006), pp. 125–134.

[61] ——, *Real-time, large scale optimization of water network systems using a subdomain approach*, Second CSRI Conference on PDE-Constrained Optimization, 2006.

[62] J. P. LEBACQUE, *Les modèles macroscopiques du traffic*, Ann. Ponts Chaussées, 67 (1993), pp. 24–45.

[63] R. J. LEVEQUE, *Finite Volume Methods for Hyperbolic Problems*, Cambridge University Press, Cambridge, New York, 2002.

[64] E. MOSEKILDE AND E. R. LARSEN, *Deterministic chaos in the beer production-distribution model*, System Dynamics Rev., 4 (1988), pp. 131–147.

[65] T. NAGATANI AND D. HELBING, *Stability analysis and stabilization strategies for linear supply chains*, Phys. A, 335 (2004), pp. 644–660.

[66] G. L. NEMHAUSER AND L. A. WOLSEY, *Integer and Combinatorial Optimization*, John Wiley & Sons, New York, Chichester, Weinheim, 1999.

[67] G. F. NEWELL, *A simplified theory of kinematic waves in highway traffic*, Trans. Res. B, 27 (1993), pp. 281–313.

[68] R. OLFATI-SABER AND R.-M. MURRAY, *Consensus problems in networks of agents with switching topology and time-delays*, IEEE Trans. Automat. Control, 9 (2004), pp. 1520–1533.

[69] Y. POCHET AND L. A. WOLSEY, *Production Planning by Mixed Integer Programming*, Springer Series in Operations Research and Financial Engineering, Springer, New York, 2006.

[70] J. SMOLLER, *Shock Waves and Reaction-Diffusion Equations*, Springer, Heidelberg, 1983.

[71] P. SPELLUCCI, *Numerische Verfahren der nichtlinearen Optimierung*, Birkhäuser Verlag, Basel, Boston, Berlin, 1993.

[72] J. D. STERMAN, *Business Dynamics*, McGraw–Hill, Boston, 2000.

[73] F. TRÖLTZSCH, *Optimalsteuerung bei partiellen differentialgleichungen*, Lecture Notes, Technische Universität Berlin, 2002.

[74] S. VOSS AND D. L. WOODRUFF, *Introduction to Computational Optimization Models for Production Planning in a Supply Chain*, 2nd ed., Springer, Berlin, 2006.

Index